High Performance Marine Vessels

Liang Yun · Alan Bliault

High Performance Marine Vessels

Springer

Liang Yun
Marine Design and Research
Institute of China
Shanghai, China

Alan Bliault
A.S Norske Shell
Sola, Norway

ISBN 978-1-4614-0868-0 e-ISBN 978-1-4614-0869-7
DOI 10.1007/978-1-4614-0869-7
Springer New York Dordrecht Heidelberg London

Library of Congress Control Number: 2012932303

Printed on acid-free paper

Springer is part of Springer Science+Business Media (www.springer.com)

Preface

Speed is not simply about velocity in air or water but should be considered in context with its purpose and the tools available. Until recently in historical terms, the motive power available for travel over the water was manpower itself or wind. Over many centuries sailing vessels have been refined so that they could harness more of its power efficiently, and reach higher into an oncoming wind so as to perform a more direct route to the objective. The wind is not available to order nevertheless, and so "speed" achieved is not constant.

The invention of reciprocating engines, initially steam driven, made a step change for maritime transport, just as it did on land a little over two centuries ago. It changed the meaning of speed over water, since not only could a vessel be designed to travel directly to its destination, but also could transport much greater payloads than possible previously, and could deliver independent of the weather environment.

In the first century of powered marine craft, speeds increased from around 20 knots to about double that. At such speed, there are challenges even for large craft due to rapid increase of drag on the hull if a boat continues to try to push its way through. The propeller driving such a vessel also loses efficiency due to a phenomenon known as cavitation unless specially designed to harness it.

In the early part of the twentieth century, pioneering engineers conquered both problems and developed planing craft that could travel much faster by skimming over the water surface. The racing fraternity that grew in this period took things to the limit and produced craft that were in danger of flying if a stray gust of wind should hit. Commercial and military craft have not tested these boundaries quite so close, even though in the last half century service speeds for passenger ferry transport have doubled.

The search for more speed—humanity has a tendency always to seek more—has been enabled through increasingly efficient and lightweight power plants such as high-speed diesels and gas turbine engines, and lighter and stronger structural materials (aluminium alloy, GRP, titanium alloy), that have enabled designers of fast boats, hydrofoils, and air-cushion craft to develop performance close to physical limits of speed in a seaway.

In the last 30 years or so, a revolution in electronics has given us the possibility for automated stabilization of motions that was simply not possible before, together with big strides in power plant efficiency, not to mention satellite navigation. These have been important enablers to comfort at higher speed and high-speed vessel development.

A series of new variations of high-speed craft or high-performance marine vessels (HPMV) have been developed in the last half century, including improved planing monohull craft from the 1940s, hydrofoils from the 1950s air-cushion vehicles and surface effect ships from the 1960s, small water plane area twin hull craft from the 1970s, high-speed catamarans from the 1980s, wave piercing craft from the 1990s, high-speed trimarans in the first decade of twenty-first century, and wing in ground effect craft from 1970s to the present. These various concepts and the hybrids that we will describe form an interacting group of vehicle concepts.

Designers, scientists, and various organizations both commercial, military and governmental have dedicated resources over the last century, and particularly heavily in the last 50 years to find ways that combinations of the hull geometries, hydrofoils, and static or dynamic air cushions can be used to deliver speedy vessels that can perform very challenging missions. This work continues, still strongly driven by military objectives, and increasingly now by energy efficiency and environmental impact rather than simply the mission envelope defined by speed/payload/environment/range.

A product such as a high-speed marine vessel can only be successful if it is able to fulfil a market requirement in an appropriate way. To deliver people or cargo efficiently, there must be a timing fit, often with other transportation linking in at each end of the mission. This applies in the military environment just as much as for Ferries or utility missions. As the other transport elements develop, this also changes the demand for the marine transport and can affect their continued success in service. Until recently it has been the cost of fuel that has played a large part in HPMV economy. While this continues, the cost inclusive of environmental impact is becoming a strong driver to further develop powering efficiency.

Both technology and human society are continually developing. To the present largely fuelled by hydrocarbon-derived energy, this societal development has accelerated greatly over the past half century as the population has also grown worldwide mainly concentrated around large cities. Fast marine craft have matured, while still having a much wider variety of concepts in use than that available for passenger aircraft that have converged to a narrow variation around a geometrical configuration and mass production. Maybe this is because the range of applications is much wider for HPMV. It does at least continue to offer opportunities for Aero-Marine Engineers to be involved in a wide range of concepts and challenging operations!

In this book, we refer to the craft family as HPMV, as the vessels are not only built for high speed, but may also have other attributes such as amphibious capability (air-cushion vehicles) or extreme seaworthiness (SWATH). Specialists from some countries refer to such craft simply as high-speed craft (HSC); however, the use of HPMV is more common now, and we will use that description and acronym in this book.

The authors have been concerned with HPMV for a long time. Liang Yun has more than 40 years of experience at the Marine Design and Research Institute of China, Shanghai (MARIC), and 20 years plus as the Chairman of HPMV Design subcommittee of the China Society of Naval Architecture and Marine engineering, CSNAME, as well as organizing the annual International HPMV Conference, Shanghai, China for a dozen years. He has been involved in ACV development in China since the very first prototypes were constructed in Harbin in the late 1950s and has been involved to some extent in design of many other vessel types treated here. Alan Bliault has also worked in the ACV industry in its early days as a Naval Architect, but became involved in the offshore oil industry in the early 1980s and so has led a double life since that time, in order to maintain his connections with the world of fast marine craft.

Some while ago, we decided to write a series of books on the analysis and design of different HPMV and have completed two on individual craft: "Theory and Design of Air Cushion Craft" (2000); "WIG and Ekranoplan, Ground Effect Technology" (2009) presenting the hydrodynamic and aerodynamic theory behind these two types. This will be followed by volumes on Catamarans/Trimarans and Monohulls/Hydrofoils in due course.

We do feel that many people have a strong interest in this technology, however while many HPMV are in operation in different parts of the world, until now there has not been a single volume giving an overview, discussing the differences and special features between them, as well as the approach to selection taken in various cases for both civil and military applications. So, we present this book entitled "High-Performance Marine Vessels" for reader's interest.

We cover as many HPMV concepts as practical within a single volume, with technical summary descriptions and discussion of the design drivers as an introduction for a wide readership. We include many pictures and figures describing the shapes and configurations as well as features of various HPMV, together with some tables to introduce the leading particulars of the craft types. Our idea with this book has been to survey HPMV development, the market drivers, and the responses over time of the marine construction industry.

The book introduces the HPMV family of craft concepts in Chap. 1. Chapters 2–6 introduce successively the ACV, SES, WIG, Hydrofoils, Monohulls, Catamarans, Wave piercers, and Trimarans. In Chap. 7 hybrid and novel HPMV configurations are surveyed. A review and comparison of various HPMV, and the strong competition between various types in the worldwide civil and military markets through their development and their future prospects is covered in Chap. 8.

We have included an appendix summarizing the British "InterService Hovercraft Trials Unit" IHTU, and its successor NHTU as an example of how the military development provided leverage for a concept development. In the USA, the hydrofoil was supported through a series of programmes, and in Russia the ekranoplan followed such a programme. Both are referred to in the main chapters and detail is available in the references and resources.

If the reader is encouraged by this book to dig deeper, then at the end of this book there are references used and a listing of sources that are useful in enquiry into

HPMV. The internet can be rather a maze, so we hope this listing can be a help to home in on useful data without getting too sidetracked.

The book is written particularly for the following readers:

- Students in middle and high schools
- Students and teachers in Naval Architecture and Marine Engineering, and other concerned faculties of universities and institutes
- Staff members, technicians, and engineers of marine transportation units, shipping companies, shipyards, ship research institutes, and other concerned units for both civil and naval organizations
- All people who are interested in the HPMV in both military and civil applications

In tracing the historical background to the different craft, we refer to many famous people in the marine world who have played their part in major technological achievements. There are many more than just the names you will meet here, and we salute all those who have dedicated their lives to these great endeavours. The results can bring great satisfaction, at least for a while, until the next challenge appears, and the urge to take the next step becomes imperative. There are of course disagreements in some areas as to who came up with ideas first or had the greatest impact. Readers of the Wikipedia can experience this simply by browsing the discussions behind many entries. In fact we believe it is amazing that such similar ideas can often surface at opposite sides of the world at similar times, and hope that both or all sides can celebrate the ideas themselves.

There is still a long way to go before technology development reaches its absolute limits. We are close regarding pure speed, but economy, passenger comfort, and environmental impact are still significant challenges for the twenty-first century to grapple with. We will discuss a little of this, and hope some who read this book will play their part in driving HPMV forward in the future!

The content of the book may seem in a slightly random sequence, dealing with ACV and WIG craft first. The logic is that these concepts are more out on the edges of the technology while the monohull, hydrofoil, multihull, and then the hybrids do follow a more natural sequence and interaction, allowing us to move more smoothly to a concluding chapter on opportunities. Readers may equally well start at Chap. 4 and move forward, returning to Chaps. 2 and 3 later if they wish.

Shanghai, China Liang Yun
Sola, Norway Alan Bliault

Acknowledgements

The authors would like to express their sincere thanks to the leadership and colleagues of the Marine Design and Research Institute of China (MARIC), and China Society of Naval Architecture and Marine Engineering (CSNAME), as well as Shanghai Association of Shipbuilding Industry (SASI), including Prof. Xing Wen-Hua (Managing Director, MARIC), Mr. Wang Gao Zhong (Deputy Managing Director of MARIC), Prof. Huang Ping-Tao, President, CSNAME, Mr. Zhou Zhen-Bai, President, SASI, and Mr. Yang Xin-Fa, Secretary General, SASI; Senior Engineers Sun Yong-Hua, Tao Ping-Ping, and Wu Chen-Ji, all of MARIC; Senior Information Engineer Chiao Yun from MARIC; and Prof. Liang Qi Kang, Former Managing Director of MARIC.

Thanks also to Jeffrey Hu (computer software engineer) and Kelvin X. Yun (financial specialist) for their help during the writing of this book.

Image Acknowledgements

We give special thanks to the following for their kind agreement to use their images (figure references in parentheses): The National Gallery, London (1.2); Discovery Museum, Tyne and Wear (1.5 and 1.7); Airlift Hovercraft (2.67); Supramar (3.30, 3.31, 5.4, 5.15, 5.16); Sea Eagle (3.32a, b); C&S AMT, Bang Wan Chun (3.34a, b); Wingship Technology Corporation, Jin H Park (3.35); Wigetworks (3.36); Universal Hovercraft (3.37); The Trireme Trust UK (c)1988 Paul Lipke (4.1) Arnesen Propulsion Systems (4.13); Lurrsen Shipyards, Klaus Jordan (4.16); Windy Yachts, Knut Heiberg Andersen (4.19), Millenium Superyachts (4.17); Austal (4.23, 6.4, 6.9, 6.11, 6.24–6.29, 8.25); Incat (4.24, 6.16–6.19, 8.9); Rodriquez (5.17, 5.18, 5.33, 6.10, 8.28); Condor (6.14 and 6.15); ESI/SESEU (7.6a, b, 7.7a–c); Navatek (7.15a, b); Quinetiq (7.18, 7.19, 7.20, 7.21); M Ship Company (7.22–7.27); Textron Land & Marine Systems (8.6 and 8.7); Sabdes Design, Scott Blee (8.27); NaplesImage.com, Jay Nichols (6.13); Turbine Marine (4.15); Bluebird Supporters Club, Mike Varndell (6.21 and 6.22); BMT Nigel Gee (6.30 and 6.31); Stena Line

(6.7–6.9, 6.20); British Power Boat Racing Club and Graham Stevens (1.11 and 4.14); Boeing (3.29).

We would like to thank John Gifford and the personnel of GriffonHoverwork for their help with many photos and updated information on amphibious hovercraft in Chap. 2.

Our thanks also to Alan Blunden of Fast Ferry International for images 8.10–8.14 taken on one of his many photographic forays to Hong Kong. Most appreciated. Special thanks also to Brian Russell, who contributed Appendix 1 giving a taste of the experience gained in the UK through the Interservices and Naval Hovercraft Trials Units.

We are proud to be able to present publicly available images from the US Navy and Marines throughout this book, many taken by enthusiastic crew of the various vessels and the LCAC Craft. Many missions of mercy have been undertaken by these personnel in the recent past following environmental calamities all over the world and deserve congratulation for their dedication to this work.

We have used a number of publicly available images from WIKI Commons and thank the organization for their work. The WIKI is an invaluable source of information which is recommended, so long as you take care in interpreting the information as it is constantly being updated!

Finally heartfelt thanks to our families, who have been very patient over the last 3 years while this book has been in preparation, and to the editorial staff at Springer who have been consistently encouraging, as well as incredibly patient while the final product took form.

Liang Yun
Alan Bliault

Contents

List of Figures

Chapter 1
Introduction: High-Performance Marine Vessels

Our Subject

In this book, we introduce all types of high-performance marine vessels (HPMV), aiming to familiarize the reader with the different types, how they are configured and why, the background to their design and performance including their historical background, and where the future prospects of different concepts are developing just now. This is not a textbook detailing hydrodynamics, though we do touch upon some fundamentals. Our aim is to widen awareness of the options available to deliver efficient high-speed marine transportation, whether for commercial or military purposes or for pleasure.

We make reference within the chapters to specific papers and textbooks that allow the reader to dig deeper if they wish and might otherwise be difficult to locate. At the back of the book, we include a listing of more general reference material including books, journals, and Internet locations that can form the start of a search. The design and performance of HPMV involves both hydrodynamics and aerodynamics, so the reader will find references from both of these fields.

The Excitement of Speed

From as far back as the 1700s, inventors have developed boat concepts to go faster across the water using wind as the power source, and from about the same time experimented with flying machines to conquer the air. With the advent of mechanical power the independence from using wind as a primary driving force gave a freedom not available to mariners before. As the development path progressed for marine vehicles, inventors realized that to go faster it was better to try to skim over water rather than push through it. There were proposals to blow air under dish shaped boats, and once powered airplanes had become a reality, floats were added so they could take off from water, offering a simpler take-off runway than rough fields; at least when the water was calm. While there were many failures with vehicles

L. Yun and A. Bliault, *High Performance Marine Vessels*,
DOI 10.1007/978-1-4614-0869-7_1,

that were built by so many "amateurs" in the early 1900s, the successful craft gave their inventors an amazing experience when they did skim successfully over the water or took off into flight.

There is something magical about high speed over water, particularly smooth water. In the middle of a large estuary or lake you begin to feel suspended in motion, and just the spray in your face tells you that you are speeding along. There is this feeling in a hovercraft, as well as in fast boats and small Wing in ground effect (WIG) craft. In small waves it is very different, like riding over cobblestones. This effect is felt in boats, and even in very light air cushion vehicles (ACVs). Only WIG craft avoid this as they are insulated by their flight height. In rougher seas the experience is different for each type of craft. Catamarans and hovercraft have a very different feel to a mono-hull. From the personal point of view they are much less nauseous or sick-making as the vibrations you feel are at higher frequency, and there is less rolling to disturb the human balance mechanism in the middle ear.

The Search for Performance

What is high performance? The meaning has changed with time, and could be equally applied to wind powered vessels as to machine powered craft. Speed is an important factor, though normally it is combined with ability to traverse waves as smoothly as possible. Our subject in this book is marine vessels that are able to travel at high speed. This mainly applies to engine driven vessels, though in the nineteenth century the introduction of engine powered craft was actually to enable them to become less dependent on the strength and direction of the wind to plan their voyages. It was a means of reducing the journey time by going directly towards the destination rather than taking a zigzag route as sailing vessels have to do most of the time. It is important to note that wind driven craft, particularly multihull yachts have made amazing advances in speed in the last few decades. We will not treat this subject here as it is easier considered a separated stream of technology development. The achievements have been quite spectacular though as can be seen from this picture of a trimaran hydrofoil, Fig. 1.1.

How have HPMV come about? The impetus has continually been the market pull for over water transportation at higher and higher speed. The pull has alternated between military and commercial applications, but has continued from the beginnings of motorized boats and ships. Aside from commercial transport, the attraction of using mechanical horsepower to drive craft that could travel fast has been a strong one for inventors, rich industrialists, and the aristocracy. Persons or families of significance in Europe and America in the nineteenth century would have a yacht built for their pleasure, and once mechanical power became available this was clearly the next step in convenience. Back in the early 1800s steam engines had begun to be installed in large sailing ships driving paddle wheels, and generally used at the ends of a voyage. As steam engines became smaller and more efficient they were able to be installed on smaller vessels. One useful application was as a tug to manoeuvre

Fig. 1.1 High-speed sailing hydrofoil trimaran

Fig. 1.2 A steam paddle tug bringing in HMS invincible for final break-up

larger vessels in and around ports rather than use small rowing boats. An example is shown in Turner's painting "The Fighting Temeraire", Fig. 1.2, from that era, the tug pulling the battleship "Invincible" that fought alongside the Victory at Trafalgar [1-1].

Fig. 1.3 HMY Victoria and Albert II

At this time high speed was anything above 10 knots. Clippers and some other high speed merchantmen were able to achieve speeds in the high teens, but most sailing ships and boats were much slower. In the initial decades of the nineteenth century Yachting Clubs already existed in Europe and the USA, running annual regattas for their members, and encouraging competition and the design development that was necessary to win. In England the Royal Yacht Squadron has played a key role in the development of yachts, and also motor yachts. Initially motor yachts were banned from competition in their regattas but in the late 1820s the Northern Yacht Club in Scotland set up a challenge for steam yachts at its annual regatta on the Clyde and this initiated the development of the type [1-2]. The first was a boat built for a Mr Assheton Smith, 120 ft long, and able to travel at up to 16 knots.

The British Royal Family had several paddle steam yachts built, starting with the "Victoria and Albert" in 1843 at 200 ft in length. Figure 1.3 shows their second yacht, a paddle steamer, while Fig. 1.4 shows the third vessel built late in the nineteenth century which had screw propulsion. The development of these vessels followed a trend of providing high quality accommodation, and a comfortable ride for the voyages that might be envisaged. For European aristocracy this was mostly day pleasure cruises, and perhaps short coastal trips while by the end of the nineteenth century it had moved on to transporting VIP's and their entourage for foreign official visits. Figure 1.4 shows HMS Victoria and Albert visiting Christiania (now Oslo) in Norway. It was not until mid twentieth century that these vessels began to be built for global voyages, and mirrored the development of smaller passenger liners, prior to the advent of mass air travel.

These types of vessels have continued to develop as they are still attractive to the rich and famous! In the mid twentieth century there was a split between craft built

Fig. 1.4 HMY Victoria and Albert III

for wealthy clients focused on the excitement of speed; and slower long distance vessels which could be stationed in the Mediterranean for a period for cruising, and then taken across the Atlantic for a cruising period in the Caribbean. The use of speed in the cruising locations is essentially to be able to take the owners from port to a location of beauty or seclusion, and later to return with the exhilaration of high speed again.

Going back to the 1880s, in 1882 the reaction steam turbine was invented by Sir Charles Parsons in Great Britain and shortly thereafter he had the 100 foot long demonstration craft "Turbinia" built, Fig. 1.5. She could travel at 34 knots powered by her 2,000 hp turbine providing a very successful demonstration of the steam turbine motor, that lead to much business for onshore industrial applications. The British Royal Navy was a rather more difficult customer to capture and this led to Turbinia being navigated at high speed through the assembled fleet at Spithead off the Isle of Wight on the occasion of Queen Victoria's Diamond Jubilee in 1897, causing considerable disturbance, and (no doubt) the advertising Sir Charles intended. Turbinia is now located in the maritime museum in Newcastle, England, see resources at the end of this book for location details.

At that time it was already practical to design and build "gun boats" powered by reciprocating steam engines that could take control of remote coastal/estuarine/river locations in the British Empire, and the Royal Navy took advantage of this to procure a significant number of such craft and deploy them as far away from England as the Yangtze river in China and in South America.

At the same time as Parsons introduced his turbine, oil burning internal combustion engines were invented and the motor car became a reality. In 1886 Gotleib Daimler put a 1.5 hp benzine engine in a 24 foot launch to demonstrate its use. This was the start of a new trend where private owners built boats with larger and larger engines aimed at higher speeds. Speeds rose through 20 knots at the turn of the

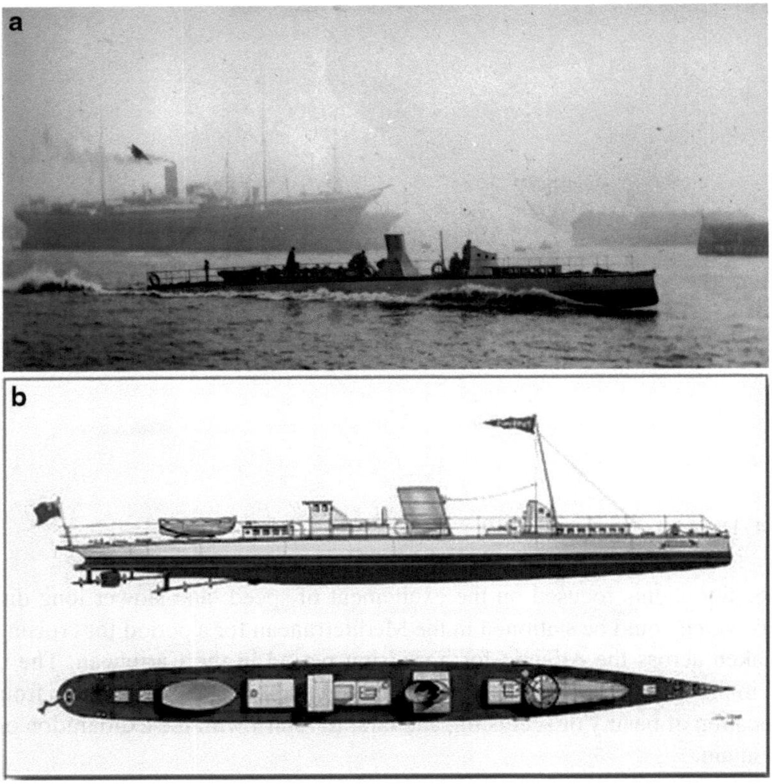

Fig. 1.5 (a) Sir Charles Parson's Turbinia at speed. (b) A plan and side view of the Turbinia

twentieth century up though 30 knots at the end of the first decade. The First World War saw the rapid development of engines for aircraft, and this gave an opportunity for speed boat enthusiasts shortly thereafter. Offshore races were organized in this period, further encouraging development. Engines of 800 and up to several thousand horsepower were installed in craft that achieved speeds above 100 mph.

The Second World War brought light weight engines such as the Rolls Royce Merlin engine used in the famous spitfire fighter aircraft, a compact unit giving around 2,500 hp. When installed in a speed boat for Donald Campbell in the United Kingdom, he was able to raise the world water speed record to 130 mph on Lake Maggiore in Switzerland. The hull form to safely achieve this had already departed from the planing mono-hull to a three pointed hydroplane.

Boats built for high speed competition have become a specialized design approach, and one that is optimized for the particular challenge. From the 1940s competition over deep water coastal courses has driven advances with the deep V monohull, while racing around smaller closed courses has driven development of the hydroplane; initially multi-stepped single hulls with shallow V bottom hull form, and later advancing to catamaran configuration with stepped side hulls and

Fig. 1.6 HMS Warrior

aerofoil shaped centre structure. The absolute speed record boats moved from large reciprocating engine and propeller drive to aircraft jet propulsion in the 1950s, while offshore racing boats have moved to surface riding propellers and gas turbine propulsion in the unlimited race class. Storage of fuel for endurance is not a controlling factor for racing craft and so power and its delivery for propulsion controls how close to the limit you can get. The limit is defined by craft stability in waves to avoid being pitched up too far. In the case of the speed record boats the waves of concern are just ripples on the surface.

If we go back to the mid 1800s when steam propulsion was being introduced into sailing ships to increase their service speed, military clients supported naval architects and shipbuilders to experiment with helical screw propulsion as an improvement on the paddle wheel. It is interesting to consider the contradiction between the paddle wheels needing an upright ship, and the sails that will always give the ship a roll to one side or the other. If the ship's sails are all furled, we only have the motion due to the waves to contend with. If the mechanical power is used for leaving and entering port we have a match. Out in the ocean the paddles are actually a drag unless allowed to rotate, a serious problem for paddle steamers with wheels amidships on each side.

The screw propeller was less sensitive to waves passing the ship, and to the ship's rolling. Initially screw propellers were relatively large diameter and rotated slowly, driven by reciprocating steam engines mounted just aft of amidships. An example of this may be seen in HMS Warrior, one of the first screw driven steam and sail warships for the British Royal Navy built in 1860, Fig. 1.6. The Warrior can be visited in Portsmouth, England, as it has been restored and is on display as a working museum.

Fig. 1.7 A Stern view showing the three propellers on each of three shafts of Turbinia

The demand for increasing power led to development of steam turbine propulsion, initially pulled forward by demands of the military market leading up to the First World War in the early twentieth century, and later by the transatlantic passenger liners that competed for the "Blue Riband" for fastest time between Europe and New York. Oil displaced coal as the fuel for combustion, as it was more compact to store in tanks in the ship and could be pumped to the burners. Turbines were initially large and heavy, suited to very large ships, but not so well to gunboats, and so these smaller vessels moved to using the diesel engine from reciprocating steam power. Slow and medium speed diesels gave way to high-speed diesels for small high speed craft such as the patrol boat which was developed and perfected by the Second World War as a rapid deployment defence against submarines.

Patrol Boats had a different hull form from the rounded shape of slower vessels. Their hull had a flat canted bottom, and in some cases a stepped hull to minimize drag when the craft was "planing" at high speed. These monohull vessels were fast—speeds as high as 50 knots were achieved, but the power to achieve this was prodigious. While in the 1940s patrol boats were powered by large reciprocating motors, once the gas turbine had been invented and developed for use in airplanes, it was also installed in patrol boats to take the speed up to the limit. Prior to this they borrowed from airplane technology using reciprocating aero engines as a means of supplying lightweight power, even though this meant that their range was relatively small.

The propellers used to drive these fast craft evolved from the large slow speed designs to much smaller propellers with spoon shaped blades, and later more like the blades on a turbine. The Turbinia is an interesting example. Here, Parsons placed several propellers in series along the three canted shafts, Fig. 1.7. By doing this he was able to keep the propeller rotational speed low, avoid "cavitation" and so be able to deploy the available power efficiently. It took some decades before naval architects perfected design of propellers to work well with the cavitation phenomenon, see refs. [1-3, 1-4].

In a seaway the ride in a planing boat can be very uncomfortable for the crew. In a choppy seaway the waves give a ride rather like a horse drawn carriage without spring suspension riding over cobblestones. In heavier seas, a boat can take off from the water rather like a porpoise jumping. To use these craft as passenger vessels, the service speed had to be accepted as lower, in the range of 20–30 knots. At these speeds the waves still cause pressure pulses, but not at the jarring force of higher speed. In the first half of the twentieth century many different designs were built, gradually improving efficiency by attention to the propulsion system design, and matching this to the hull form. Hull forms were also developed with steeper canted planing surfaces, and "stepped" forms which have also proved best for seaplane floats. In the last 50 years, designers have continued to experiment with catamaran hull forms, and more recently with the use of water jets for propulsion rather than open water screw propulsion.

In the speed range up to 30 knots, where planing lift forces are limited it may be considered more efficient to optimize the underwater form of the vessel so as to reduce drag forces and vertical motion response. This leads to very long thin vessels having limited inherent roll stability. One way to overcome this is to have a submarine like lower hull and minimized water plane. The idea was found to work best if a catamaran form was used, and this lead to development of SWATH (small water plane twin hull) vessels that have proved useful for these medium speeds in open ocean wave conditions. Pressing the concept forward for higher speeds in the range above 35 knots leads us to the wave piercing catamaran. This concept also has a relatively narrow speed range (Froude number range) where it is optimum, above which powering requirements rapidly increase. The concept has to be successful enough commercially to support scaling up to very large size, and this has been a driving force also for fast ferry developments in the last 30 years.

Single hull craft were able to ride safely at speeds up to 80 knots or so once the stepped forms were developed so that the hull effectively ran on a number of discrete small supporting surfaces. A catamaran form helps with transverse stability, while at higher speeds it was found more stable to use a central hull and two forward floats called sponsons giving a three-point support. This was adopted by the Campbell family in the UK in their water speed record boats in the 1930s–1960s.

A move in the opposite direction is taken by builders of "airboats" that are used in the everglades of Florida. Basically, an automobile or light aircraft engine is installed at the stern of a shallow flat bottomed barge like hull driving a large air propeller, Fig. 1.8. These boats can travel at speeds up to 50 knots over very shallow water. A very exciting ride, but they are not suited to open waters though! They are used for hunting, fishing, and tourist trips and have developed since the 1950s.

We have focused here on the central thread of the monohull story. From the beginning of the twentieth century pioneers such as Forlanini in Italy experimented with hydrofoil supported boats in an effort to separate the hull from the water surface and so increase speed. This work was moved forward by Baron Von Schertel in

Fig. 1.8 Airboats at speed

Switzerland, and Alexeev in Russia, realizing commercial passenger craft. Alexeyev's river skimmers were built in significant numbers on production lines to help speed communications along Russia's river arteries in the 1950s–1980s, while von Schertel's seagoing hydrofoils changed the face of many coastal passenger connections in Scandinavia and the Mediterranean through the same period. More of this will be described in Chap. 5. First, if we are to go faster, we must reduce drag while maintaining a stable craft. That leads us to look at the options available.

Resistance to Motion: Drag in Water and Air

Airplanes fly in air with a density only 1/800 of water, and so with much smaller resistance to motion than a boat. Only when an airplane reaches close to the velocity of sound (the "sound barrier" at 1,250 kph) does drag start to increase at a significantly higher rate due to generation of pressure waves at the nose and wing leading edges.

The weight of an automobile is supported by its wheels. When a car runs on a smooth surface with well inflated tyre, rolling resistance increases very slowly. The other components of resistance to motion are the air resistance of the body shape, and the resistance due to rotating wheels and transmission shafts in their bearings. These also increase rather slowly with forward speed. Air drag is the dominating factor, particularly for fast cars that can run at more than 160 kph, though the majority of the power installed in a car is to enable it to accelerate from standstill.

Ships and boats are rather different from aircraft and automobiles. The ship weight when still is supported in accordance with Archimedes principle that "the buoyant force on the ship at the water surface is equal to the weight of water displaced by the ship below the waterline". When a ship starts to move, its hull below the water line has to move through the water rather like a car moving through air. This holds true until the speed of the boat can generate a vertical reaction force from the water, "dynamic pressure" to start to lift it. When this dynamic pressure force can balance the weight of the boat then it is said to be "planing".

The resistance components of a ship moving in water include the hydrodynamic friction and vortex resistance from water wetting the ship's immersed surfaces, and also include another component which can be a large part of resistance, created by waves generated by the ship hull motion in the water. The ship generates a pressure field all around the hull which radiates outwards and creates a regular pattern of surface waves around it. The energy needed to do this is what creates the wave drag force.

Water friction is also much higher than that of air on a similar surface area since the boundary layer—the layer of fluid where the local velocity reduces from free stream down to zero at the hull surface—is in a fluid rather than a gas and the energy to force the molecules to move relative to one another against the attraction forces is so much greater.

Designing boats for high speed is one of either installing lots of power, or finding a way to minimize the contact a boat has with the water surface. As we have heard above, as engines of higher power and smaller size have become available, they have very soon been installed in boats, and the potential for higher speed has been demonstrated. Operational range has been a more challenging target to attain, and has been enabled by constant improvement in engine efficiency and improving liquid fuels. In the last half century, speed boats have reached their logical limits, rather like airplanes approaching the speed of sound, so the technology challenges have been related to design of craft that can give comfort as well as high speed, at least in the range of 30–60 knots, whether for competition, passenger transport, or military purposes. The starting point for our investigation of the different vessel types is the understanding of take-off or planing as a phenomenon.

Reducing Resistance to Motion

There are two main running modes when operating in and on the water. A craft at slow speed operates in displacement mode, while at high speed it can use dynamic forces to slide or "plane" on the surface rather than pushing through it, or it can use an air cushion to lift it from the water surface. Alternatives to the air cushion to lift a hull out of the water are to use foils in the water attached to struts (the hydrofoil) or add wings that support the weight above a take-off speed (WIG craft).

In between these two modes, there is a transition that is characterized by a peak in the drag curve. For ACVs or surface effect ships (SES) this is called "hump speed". Monohulls or planing catamarans experience it as the point where planing starts while for a WIG craft it is just before take-off into flight. Let us consider the different modes, and discuss the resulting forces on a hull.

Fig. 1.9 Catamaran ferry at speed between Stavanger and the islands

Static Buoyancy

When running at low speed the weight of a craft is wholly supported by water displacement, and the wave-making resistance will be the principal component of drag.

Wave-making is much reduced for very fine (long and thin) hull shapes, so the best way to reduce resistance is by cutting the ship along the centre line in two parts, and extend the transverse space to a definite distance between the two hulls. A centre connection structure can be made between the hulls above the water line to create a catamaran, Fig. 1.9. The design challenge for the catamaran is the structural weight of the connecting structure, since if this is very heavy, the hull volume has to be increased, eating away at the advantage gained originally. In addition, there is a challenge to fit the propulsion system, possibly including the main engines, into the slender hulls.

Refining the hull shape as above, or maybe using three parallel slender hulls (the trimaran) has its benefits up to the speed where hydrodynamic lift starts to become a significant force (above about 25 knots). Above this speed, there is the opportunity to use dynamic forces to lift the hull(s) out of the water and reduce resistance that way.

Hydrodynamic Lift, the Principle of the Planing Hull

A flat pebble can be skimmed across the water surface. Spinning it improves its performance! Even though the density of stone is greater than water, the pebble is

Fig. 1.10 Fast deep V planing craft

Fig. 1.11 Stepped hull planing craft

able to skim and even skip over the water surface for a long distance if cast strongly enough. The lower surface of the pebble has to have an "angle of attack" so that there is an upwards component of the reaction force from the water that balances its weight. Given the right angle the pebble will skim steadily along. If thrown with too high an angle, it will tend to skip. The skipping can also be seen if there are ripples on the surface.

Planing boats are designed using this same principle Fig. 1.10, a variation being many individual surfaces at an angle—the stepped hull, Fig. 1.11. The lower surface

Fig. 1.12 Surface piercing hydrofoil

is usually not flat but curved, so that the hull trim does not have to change too much as speed increases. Some craft also have trim-able flaps or flow interrupting fences attached to the transom (rear of the hull) which can be used to adjust the craft attitude at speed.

The Hydrofoil

Ships can be lifted above the water surface with aid of hydrofoils, wing-like shapes connected to the hull by struts down through the water surface. Water is 800 times denser than air, so the hydrodynamic lift of a hydrofoil wing is in similar proportion, and the foil can be relatively small, depending on the target take-off speed.

Hydrofoils come in two basic types, those with "surface piercing" foils, Fig. 1.12 and those with more like an airplane configuration, totally submerged Fig. 1.13. The first type has the advantage of being self adjustable with speed, as the vessel just rises a bit more out of the water. The fully submerged foil has to have a dynamic trimming system to adjust the angle of attack as service speed changes, so as to maintain a level "flying height". Supramar and Rodriguez are the most famous builders of the first type, while Boeing with its Jetfoil is the most famous for the second.

Fig. 1.13 Fully submerged hydrofoil

Static Air Cushion Lift

An ACV is a craft supported by a mechanically generated air cushion, which lifts the craft by air pressure acting on the bottom of the craft, Fig. 1.14. The craft plan area has to be sufficient to balance the weight against the air pressure acting upwards on it. Since the cushion lift is caused by lift fans in the craft, the craft can rise from its floating condition without forward speed. There is still a depression over water equivalent to the air pressure when at rest, and as speed is increased the craft will trim by the stern as planing speed is reached and the craft goes through a resistance "hump". Above this speed, acceleration is experienced as resistance drops away without power being increased. Resistance does start increasing as speed is increased still further, requiring more power to be applied.

The ACV has evolved from the original concept where the high pressure cushion was contained by a peripheral jet of even higher pressure air, creating an amazing amount of spray over water, to a vehicle with rubber impregnated fabric membrane "skirts" that minimize the escape nozzle while flexing with the waves or surface undulations.

Static air cushions are also applied to the space between catamaran hulls, with a flexible seal at bow and stern. These craft, called SES, have been built and operated successfully since the 1970s–1990s, though scaling up to large size has not proven cost-effective and so the simple catamaran has become the craft of choice for large vessels carrying vehicles as well as passengers.

Fig. 1.14 Hovertravel's BHT-180 ferry traversing the shallows at Ryde

Aerodynamic Lift

With high-speed airflow acting on the wing of an airplane, aerodynamic lift is generated so as to support the weight of the airplane. When an airplane flies close to the ground or water surface, the aerodynamic lift increases significantly, and lift induced drag also decreases. This phenomenon is called Ground Effect or Surface Effect, and craft designed to operate close to ground or water using this are called WIG craft or Ekranoplan in Russia, Fig. 1.15. The phenomenon was first discovered by Russian scientists working on hydrofoil development in the city of Gorky. The zone in which WIG operate is referred to as the surface effect zone (SEZ). WIG craft have to be designed a little differently to an aeroplane since the fuselage now needs to float on the water and plane as a fast boat before the aerodynamic forces take over and lift it off. A WIG can be thought of as a "flying boat" designed with wings to trap air for surface effect so as to reduce installed power for a given payload.

Passenger Comfort and Other Requirements

Whether people ride in the airplanes, ships, or cars, they can always feel the motion of the vehicle. The motions are due to the disturbance of the air and/or water reacting on the vehicle, transmitted through the suspension system, whether this is springs as on a car, the flexibility of an airplane's wings, or the pneumatic suspension of an ACV or SES cushion. A boat has the hardest suspension of all, simply the shape of the hull as it pushes water aside. The ability of the vehicle design to dampen

Fig. 1.15 Alexeyev SM-8 wing in ground effect (WIG)

this response, and to have a positive stability to return to the steady running attitude is a very important attribute. For aircraft this is called airworthiness, for road vehicles ride quality, and for marine vessels it is seakeeping quality, or seaworthiness.

When a ship is navigated at sea, two other features are also pursued, i.e. *manoeuvrability*, including high speed and fast turning ability, which are extremely important for the high-speed vessels in order to avoid slower moving traffic; and *safety* as well as ride comfort, i.e. the stability, safety, and comfortable ride (including rolling, pitching, and heaving) which is called seaworthiness.

Seaworthiness is a most important feature in the HPMV family in addition to the high-speed requirement, and designers make every effort to create craft with both optimized powering performance and fine seaworthiness. Apart from dynamic stability, seaworthiness can be assessed from the vessel motion and acceleration response spectra, and the human response to these. The human body is most sensitive to vertical motion at low vibration at periods between 7 and 0.7 s when nausea is induced, while physical discomfort is particularly felt at periods between 0.5 and 0.1 s cycle time. Rolling motion can also cause discomfort and disorientation at periods between 2 and 0.5 s depending on amplitude.

A WIG has a highly damped response once flying, while an ACV is highly damped in the low frequency range due to its skirt, but can transmit higher frequencies (cobble-stoning). The hydrofoil and catamaran are more responsive to the low frequencies through use of fine hull forms; and minimal rolling motion.

Reducing wave disturbance is the best way to improve a ship's seaworthiness. This generally involves designing the hull(s) to have very fine entry lines both for forward motion and also for heave. The shape of the bow half of the hull is also important to dampen pitch motion while not inducing too high accelerations.

Fig. 1.16 High-speed small water plane twin hull (SWATH), Seagull 2

Fig. 1.17 Wave piercing catamaran HSV 2

Two designs based on the catamaran are useful to give as examples of this directional approach:

- The SWATH with most buoyancy under the water surface in torpedo like hull forms, thus giving the vessel a very slender water plane so as to reduce wave generation. Figure 1.16 shows the SWATH craft "Seagull 2" from Japan.
- The Wave piercing twin hull craft (WPC) is also based on a slender water plane catamaran hull form for reducing the wave interference. Figure 1.17 shows the HSV 2 operated by the US Navy.

ACVs, SES, and Hydrofoil craft (HYC) all break free from the water surface at cruise speed to reduce wave generation. However, since ACV and SES are operating at the interface between the water and air medium, waves are still generated. Wave generation causes the hump drag phenomenon. In a seaway, amphibious ACVs propelled by air propellers tend to have more rapid speed decay in waves than water propelled HPMV, and are more efficient in light weather conditions. SES can cope with heavier seas but suffer the powering challenge similar to catamarans while requiring a rather lighter hull structure to be efficient. The seakeeping quality of the hydrofoil is rather better, particularly for fully submerged HYC, due to the hydrofoils being deeply submerged into the water. The WIG is able to fly in air, free from the wave interference other than the pressure variations from the undulating surface when cruising; however, during take-off and landing seakeeping is a significant problem.

So we may summarize that HPMV need to have some, though not all, of the following operating features to deliver the desired performance, depending on the application:

- High speed compared with other marine craft
- Seaworthiness including longitudinal, transverse, and vertical stability, and well damped motion (rolling, pitching, and heaving angle and accelerations) tuned to the service environment of speed and sea state
- Rapid turning capability for collision avoidance of slower craft
- High noise damping both internal and external due to the high installed power
- Landing and obstacle clearance capabilities for amphibious craft
- Light weight construction

The features that dominate the design specification are high speed, both in calm sea conditions and rough seas/high winds; and the lightweight structure and power plant required to achieve this. Building a marine craft that can go fast demands resistance forces are low so as to avoid high power installation with resulting high costs and low return on investment, or insufficient endurance for military or racing craft. In order to reduce the water resistance we have to make the craft clear from the water partly or completely. There are various high-speed craft configurations to generate lift to support the craft weight as we have discussed above. Such lift includes hydrodynamic lift, air cushion lift, and aerodynamic lift. If the craft speed is too low to generate sufficient dynamic lift, then a designer can use a more slender shape and form a catamaran. In this case, the buoyancy is still the main lift force supporting the ship.

The proportion of each of the three different types of "lift force" used by these craft types can be represented by a triangle diagram as shown in Fig. 1.18. The top of the triangle represents the static lift (buoyancy due to displacement), the left side of triangle represents powered lift, and right side of the triangle represents dynamic lift.

Such supporting lift components can be combined to form hybrid craft types, such as the foil-assisted catamaran (mixed foil and CAT); planing CAT (CAT plus planing hull); air cushion CAT (air cushion technology plus CAT); foil assisted air cushion twin hull (foil plus air cushion plus CAT); amphibious WIG (WIG plus

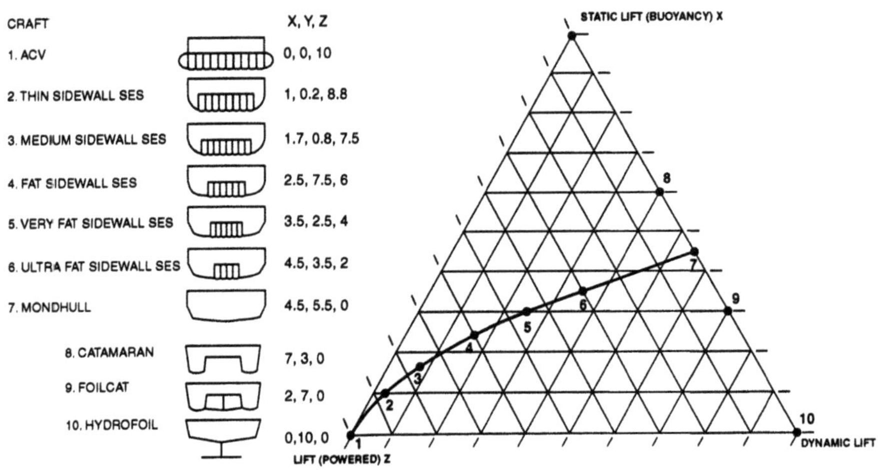

Fig. 1.18 Proportions of different lift force for high-performance marine vessels (HPMV) concepts

ACV); foil assisted small water plane area twin hull (foil plus SWATH); foil assisted wave-piercing CAT (foil plus WPCAT), etc.

The reasons for adopting combinations of concepts often has to do with the specific mission requirements, for example high seaworthiness at slower speed and higher speed in lower sea-states (foil assisted SWATH). We will discuss the merits of some of these in later chapters. Some of these hybrid craft are rather complex, and thus expensive to build with little performance gain compared to a "simple" catamaran for example, and have therefore not been used for commercial craft.

A number of craft transverse sections are shown in Fig. 1.19. In addition to the varying proportions of hydrodynamic and aerostatic lift shown in Figs. 1.18 and 1.19, we show the variations of hull form aimed at minimizing resistance while using buoyancy (craft with thin waterlines such as wave piercing catamarans, SWATH craft, trimarans and pentamarans) to give visual examples of the different craft types discussed so far, before we dive into the more detailed discussion of the different craft in the following chapters. The WIG is a special concept since as well as operating in a similar manner to a multihull planing craft, the dynamic air cushion under its wings increasingly supports it as an ACV until free air velocity is sufficient to provide support for the craft to take off and fly as a purely aerodynamic vehicle in the SEZ.

Costs: Construction and Operation

HPMV also have to be cost-effective to build, operate, and maintain. These requirements are actually different for a military craft compared with a passenger ferry or a utility craft. Ferries operate over a fixed route for which the craft design can be optimized, while military and utility craft have a much more variable environment

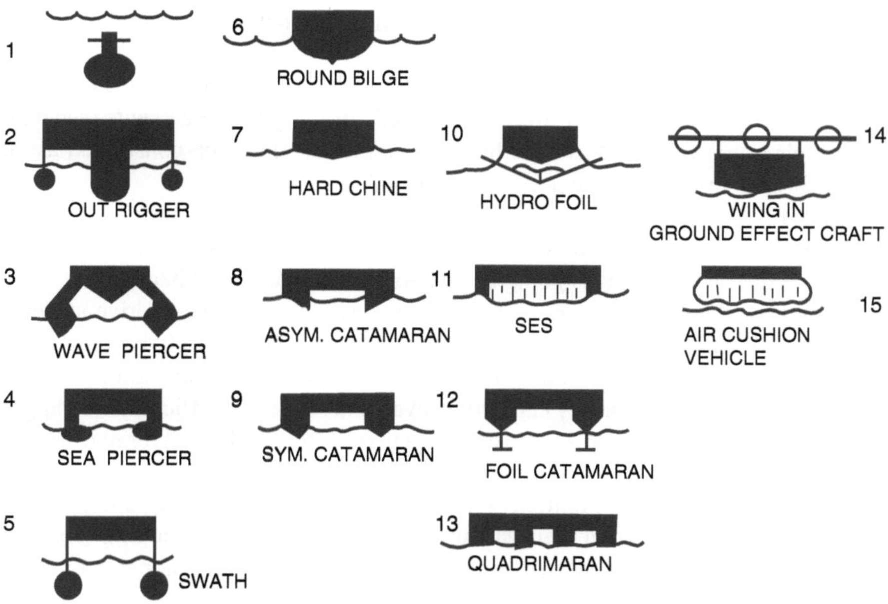

Fig. 1.19 HPMV hull section geometries

and irregular operating pattern. We discuss the relative costs and operating economy in Chap. 8, and how this has affected the development of different craft types in recent times.

Environment and the Future

Before launching into details of the main craft types, it is helpful to consider the path forward. Over the last couple of centuries, man has created mechanical machinery that can deliver the power needed to achieve almost any objective as far as transportation is concerned. Through the twentieth century, the concepts moved forward testing the barriers set by physics (hydrodynamics and aerodynamics), and we will meet a number of these as they apply to boats, hovercraft, WIG, etc. So, what are the real challenges now?

In recent years, there have been a number of economical cycles on an international scale that have initially encouraged rapid development of personal travel, both business and pleasure, and then a temporary "recession" has pulled everything back. Ferry companies operating high-speed craft have found themselves struggling to make money, and have needed to find more efficient vessels to deliver their service. Such recessions have affected the military market also, but the effects on military forces come into effect with a kind of "delayed action" due to the long timescale for delivery of a military vessel programme. Pleasure markets have reacted at the other end of the scale, so in fact in the recession 2008–2010 the pleasure boat market almost fell off a cliff in Scandinavia for example. It will recover, but may take a while.

What about wars, you will ask, surely WW1 and WW2 had a dramatic effect on technology in a short space of time? Yes indeed, and maybe these events (separate from their tragic consequences for human communities) can be considered as an accentuated investment boom and recession cycle.

Currently, the world is waking up to the linked issues of the need to move on from fossil fuels and the environmental consequences of their use, to alternative power sources. Clearly this is going to take some time, but it will happen.

Right now there are a number of motor cars available that have drive systems using a combination of electrical motors, and engines working as a hybrid to reduce emissions. Pure electric cars are currently much lower power, and limited range. Due to the link between battery capacity, power, and range it is difficult to see larger vehicles developing along this same path, whether for road or indeed for marine use. Perhaps the direct alternative will be the fuel cell, since this frees the vehicle from the limitations of batteries. At higher powers, currently aircraft are beginning to turn to "cleaner" fuels to burn in their gas turbines, and this is just a bridging step perhaps, but the step to using cleaner burning fuels (lower CO_2 emissions) is relevant to all transportation. The long term future for aircraft is unclear here, though—strange though it may seem longer term it may be more efficient to go much higher in the atmosphere for long journeys, and reduce the consequence that way. The technologies to do this are not likely to help us with marine vehicles though. The next steps for hydrofoils and WIG are likely to be delivered through developments for light aviation as this is driven by governmental policies towards lower emissions.

The use of biofuel for diesel powered craft and perhaps LNG for larger craft are two developments that are a predictable out-step from the present for HPMV. These improvements should happen "naturally" over the next couple of decades as the same technologies develop for use in road vehicles and larger ships.

Maybe in the marine competition arena we should be starting challenge events for HPMV powered by alternative energy, to drive technology development in this twenty-first century. The approach certainly worked in the nineteenth century as referred to earlier in this chapter!

While we now understand the mechanics of HPMV and have the power to take them to physical limits—and beyond if we are not careful—we are now inexorably moving into the era of alternative energy machinery that delivers power with minimum environmental impact. Short term economic cycles will inject wake up calls from time to time, while long term it is about environmental stability.

Sailing craft developed for three millennia before mechanical power changed the face of transportation in the nineteenth century AD. It has only taken two centuries for fossil fuel driven mechanical power to drive marine craft to their practical limits in terms of speed for the particular concepts. The next challenge is the "sustainable" power system, i.e. one that can have a useful lifetime at least as long as wind power!

First, we need to take a look at the range of concepts and how they were developed. We start with the ACV.

Chapter 2
Air Cushion Craft

It Started with a Coffee Tin

In England, during 1956 Christopher Cockerell assembled a coffee tin with another smaller tin soldered inside to create an annular space. He fed pressurized air from an electric hair drier attached to the assembly into the space between the tins around the periphery. The air formed a thin jet curtain around the tin, and he proved that the action would create an air cushion contained by the jet curtain sufficient to support a considerable weight, (Fig. 2.1). Later, Cockerell discovered that by directing the peripheral jet with an inclination angle inward, the jet blowing into the air cushion improved the containment, and significantly improved hovering height, potentially enabling up to 10–20 cm flying height for a full scale hovercraft. This gave the possibility of a practical air cushion vehicle, or ACV (Fig. 2.2).

After a search for sponsors, Cockerell obtained support for development of a full scale prototype from the British government, initially bound by a secrecy agreement due to the possible military potential. The prototype ACV with peripheral jet cushion was built by Saunders Roe on the Isle of Wight, and designated the SR.N1 [2-1]. At this time, the SR.N1 used air ducted from the lift fan to propel it, or stop it. The craft weighed 3.4 t, and was powered by a Leonides aviation type piston engine with total power of 320 kW; using 70% of the power for lift and 30% for propulsion. In June 1959 it started trials on the Solent clocking up 14 trial runs totaling 24 h of operation; the longest being between Cowes and Eastney beach at Portsmouth. July 1959 was the 50th anniversary of Bleriot's flight across the Channel between England and France. The Saunders Roe development team agreed after some debate to make the crossing with SR.N1. The Royal Navy was enlisted successfully to provide a Lighter vessel RNA 54 to take SR.N1 to Dover, and also provide an escort, the HRMT "Warden" for the voyage. After arrival at Dover on 24 July it was found that the weather projection for the next few days was from NE which would make the journey to Calais extremely difficult. It was decided to take SR-N1 to Calais and make the voyage in the opposite direction, so RN54 took the craft over that afternoon and moored in the harbour. That day was a holiday in France, and so the

L. Yun and A. Bliault, *High Performance Marine Vessels*,
DOI 10.1007/978-1-4614-0869-7_2, © Springer Science+Business Media, LLC 2012

Fig. 2.1 Coffee tin experiment

Fig. 2.2 SR.N1 hovering without skirts

Fig. 2.3 SR.N1 channel crossing

"Flying Saucer" attracted considerable crowds. Eventually, it was agreed after requests from officials to take the craft around to the beach and make demonstrations for the crowds there. These were very successful, though some people tried to physically test the air gap under the craft, and put both themselves and the craft in some danger. The demonstration was a success nevertheless. The crew arranged for a regroup at 0600 next morning to assess weather; some stayed in berths on the two vessels, while others went to hotels in the town. Cdr Lamb, the pilot, woke at 0300 next morning and found the weather quite calm. A reconnoitre by the escort vessel confirmed flat conditions, and so a quick decision was made to leave. Calls were made to the hotels for the other crew. The craft was fuelled and prepared, and left harbour at 04.55 making a track to the NW a little up wind, with the escort following the straight track as marker. The weather rose with the sunrise, and SR.N1 experienced a number of times falling below hump at 12 knots, and sliding down swells. The crew had to move about to trim the craft a little like a sailing boat. Eventually, after 2 h at sea the craft was able to make an entrance into Dover harbour, closing in at 30 knots and landing on the beach in front of a small gathering of enthusiasts and Press in a cloud of spray (Fig. 2.3). This first hovercraft crossing of the British Channel by the SR.N1 was at 24 km/h. It was a strong test of the technical and operations teams at Saunders Roe, and demonstrated that an air cushion craft did have significant potential.

Trials with SR.N1 continued for a number of years, while Westland Aircraft (who bought Saunders Roe just before the channel crossing) developed military and then ferry hovercraft largely based on aviation technology, with gas turbine engines and riveted high strength aluminium hull structures. The SR.N1 had skirts fitted to it, initially simply as extended jets, and later with more complex geometries. A Viper jet engine was also installed for high-speed propulsion to push the cushion system to its limits in its Mk V version which had extended bow and stern geometry simulating the SR.N2 plan form (Fig. 2.4).

Fig. 2.4 SR-N1 deep skirt and jet propulsion

The British Government sponsored in parallel an R&D company called Hovercraft Development Limited. HDL focused on responsive skirts for low resistance cushion systems, and marine based hull structures for reduced cost as an alternative to the aerospace derived technology used by Saunders Roe.

Flexible Skirts

Although the SRN1 crossed the English Channel successfully in 1959, it was with a hovering height of just 20 cm. It could only be operated over smooth ground or relatively calm sea. At that time, it was difficult to see how such an ACV could be put to practical use. It was the innovation of the flexible skirt by Mr. C.H. Latimer Needham, another British inventor, in 1957, that made it possible to increase effective air gap and bring the ACV into practical use.

Using a flexible skirt the effective hovering height for a craft of the size of SR.N1 could be increased from 20 cm to as much as 1 m, enabling a low footprint pressure ACV to negotiate rivers and rapids, estuaries and shallow water areas, sea, rough ground, soft beach sand, and also swamp and ice; places that would be impossible with any other vehicle. It is no exaggeration to say that without the skirt, to-day's practical ACV would not exist, and the skirt for ACV, was a technology enabler just as the pneumatic tyre is for a car. Figure 2.5 shows several types of early inflatable skirt, illustrating the development from rigid peripheral jet to inflated bag with short nozzle and on to the bag and finger type. The convoluted finger or segment was the

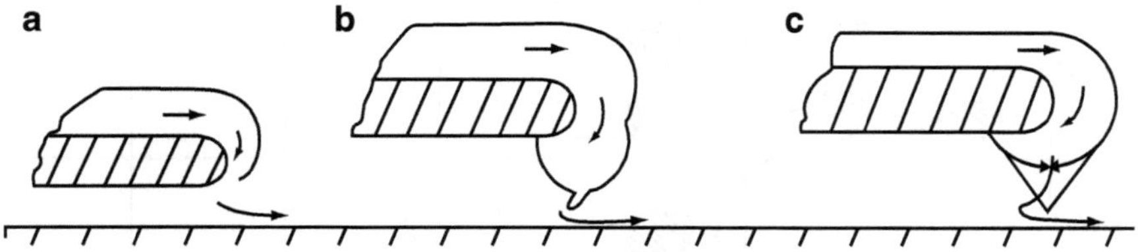

Fig. 2.5 Skirt cross section development

idea of another inventor, Dennis Bliss, who worked for the company set up to develop Cockerell's ideas, Hovercraft Development Limited.

Flexible skirts bring the following attributes to an ACV:

- Practical obstacle clearance
- True amphibious capabilities
- Reduced water resistance in waves
- Reduced impact loads on the hull from sea waves or obstacles
- Improved seaworthiness
- Improved maintainability, since the flexible skirt can be easily replaced
- Significant reduction of lift power, due to minimized requirement for clear air gap
- Reduction of water spray and particles such as sand thrown up by the craft, that caused significant problems for the engines, fans and propellers on early craft

An ACV fitted with a flexible skirt and air propulsion system can ride over both land and water, and can generally achieve:

- Speed as high as 70 knots in calm conditions
- Access to locations that are in or across seas, lake, rivers, estuaries, intertidal flats and shallow water, river rapids, swamp, ice- and snow-covered land or water that conventional vehicles cannot reach
- Transition from land to water to land via slipways or ramps, or across sand flats in estuarine and coastal areas, or ice cover in arctic environment

Figure 2.6 shows the evolution of British Hovercraft Corporation (BHC) skirt section design from peripheral jet to bag and segment. The pressurized air is blown from fans into the air cushion via the inflatable upper bag skirt, so as to have a higher pressure in the bag, giving greater stability. The upper part of the skirt is like an inflated tyre inner tube, except that the air flow is continuous and is ducted to each of the segments through holes in the fabric lower surface while being maintained at a pressure higher than that in the cushion itself. The fingers or segments are attached on the bag, and can deform individually to respond to an uneven surface. The lower tip of the skirt fingers wear by abrasion due to often touching the ground or water. Detachment from the bag and replacement during craft maintenance is relatively simple as the craft do not need to be jacked up.

BHC initially formed the "fingers" to provide an air jet channel. The construction was quite complex, requiring fairly heavy materials to form a stable geometry, and

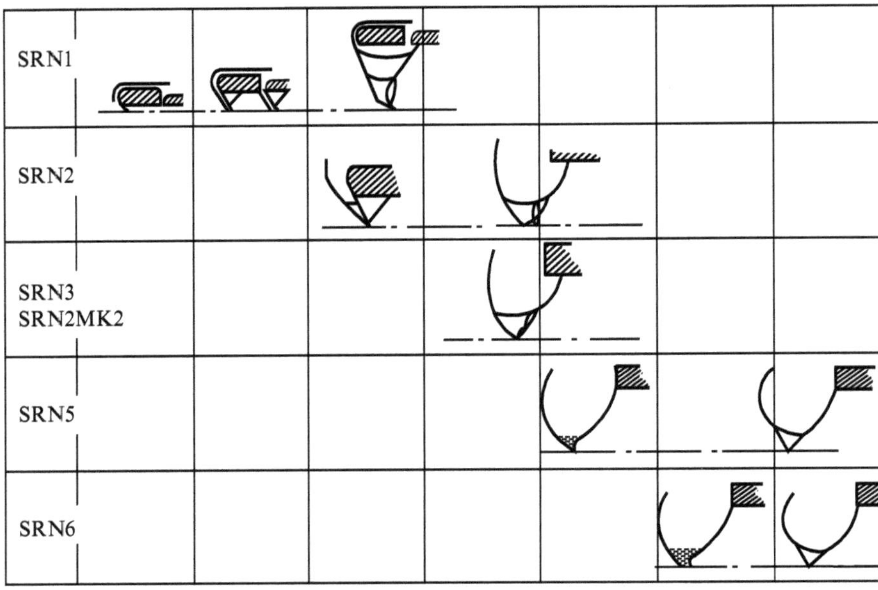

Fig. 2.6 Development of skirt designs at Saunders Roe

so the wear rate at the lower tip was somewhat high. Hovercraft Development (HDL) had a different idea to have very light weight segments and a simple plenum for the main cushion. BHC also subdivided their cushions with internal skirts to gain extra stability both fore and aft, and sideways. As the geometry of the outer bag and lower fingers improved, it was found that the internal sub-dividers could be simplified, and eventually also the fingers became similar to HDL's segments, though in heavier material. Designing a skirt is a bit like designing the suspension of a car or truck with springs and shock absorbers. It took car manufacturers a number of years in the 1960s and 1970s to refine their designs. Similarly, it took the ACV industry in the UK up to the mid 1970s to gain sufficient knowledge through experimentation with models and the full size craft before skirts really gave the optimum combination of stability, damping, and low wear rate of the fingers/segments. An example of a skirt without internal stability dividers, with a swept back bow profile to give resistance against plough-in at going down wind is shown in Fig. 2.7.

Inflatable skirt material is a woven nylon fabric with a water and wear resistant coating impregnated through it. The most common used coating is a natural rubber or neoprene material. The cushion pressure under an ACV is very small (about 1–5 kpa), and only 1/10–1/6 of the footprint pressure of a person (60 kpa), which is similar to the footprint of light tank on its tracks (40 kpa). An ACV can cross a swampy environment without leaving a track, where a person would sink deep in the mud. The wear on a skirt is therefore not heavy due to straight abrasion. Over land, it is the presence of sharp obstacles that gives significant wear, while over water the vibration of the segment tip delaminates the coating from the fabric and encourages faster wear.

Fig. 2.7 Surveyor

Responsive Skirts

Seaworthiness can be improved by using the so-called responsive skirt. Briefly, the rationale is to design the bow skirt geometry such that the bag section flexes upwards while a wave passes under without the skirt transferring a strong heave force to the bow structure, so as to reduce both pitching and heaving displacement and vertical acceleration while minimizing craft resistance in waves [see [1-4], Chap. 7]. Figure 2.8 shows the responsive skirt installed on the US Navy Landing Craft Air Cushion (LCAC). The large radius bow skirt is a prominent feature of the craft, together with the purposely designed taper towards the stern giving the craft a permanent bow up trim while the lower skirt line is level with the water. The lower attachment of the bag at the bow is aft of the upper attachment line at the gunwale, but arranged such that as a wave passes the bag will deflect upward while drawing the fingers forward adding stability in pitch. The response is soft, but stable. The bag pressure has to be high enough to avoid the bow skirt being pulled backwards within its operating envelope of speed and sea state. If this were to happen, it would cause the craft to "plough in", i.e. the skirt would increase drag so much that the craft would be pitched down and brought to a stop. As a contrast, earlier craft generally had swept back bow skirt configurations to create a "stiff" geometry and so prevent tuck under of the skirt leading to plough-in, Fig. 2.7.

The responsive skirt can give improved performance due to reduced drag in a seaway, though if the bow skirt geometry is incorrect the lower dynamic longitudinal stability can lead to lower resistance against the plough-in, the opposite of that intended. Generally small sized craft are easier to design for bow skirt stability with a more swept back geometry, using an open loop and segment skirt or a design with low bag pressure.

Fig. 2.8 LCAC Responsive skirt

Experiences from the Early Days in China

Understanding "Peak Resistance", and Skirts' Contribution to Minimisation

At the early stage of research in the late 1950s, it was thought at the time that when an ACV flies over the water, there might be no wave resistance at all, and no generation of waves such as a boat hull creates. At that time author L. Yun was a young lecturer and head of a research team working on ACV research in China's Military Engineering Academy.

His team built a manned test ACV, called "33" with a plenum type cushion. This was the first experimental ACV in China, (Fig. 2.9a, b). It was tested in 1958 at a lake nearby Harbin City, in the north east of China. They found the craft ran backward, even at maximum propeller speed, due to air flow leakage from the bow seal being larger than that from stern seal, as the craft C.G. was located behind the cushion centre of area and generated a stern trim. After adjusting the CG of the craft, it started to go ahead, however, with very slow speed, and a significant spray from the middle of the craft, since the trough from the waves made by the air cushion was located there. At that time the team realized wave-making resistance was the same for ACV's as for other boats.

The theory of water surface depression by an air cushion was developed by Newman and Poole [2-2] in the UK, and the Chinese engineers used this reference to check their craft performance. However, the calculated result was far less than the air propeller thrust on the craft "33". After observation and model tests in a towing tank, it was found that the rigid bow seal on the craft generated a significant wave pattern so as to create secondary wave-making resistance that was larger than the air

Fig. 2.9 (a) Prototype "33" on land. (b) Prototype "33" over water

cushion wave-making resistance. The seal drag was greatly reduced when a flexible skirt replaced the rigid bow seal, and the resistance peak almost disappeared due to the high cushion length/beam ratio.

Plough-In and Overturning of ACV

On 1 May 1966, first trials of the first Chinese amphibious test ACV type 711-I with flexible bag and finger skirt, were begun on the Ding Sah Lake, located at the outskirts of Shanghai [2-3]. The craft sped along as fast as 80 km/h when going down wind in light breezes. It developed a heavy bow down trim due to a sudden increase of thrust from a stern wind gust. The bow skirt became seriously tucked under, and then the craft broached, yawing beam on at high speed, and finally overturning.

There were three crew; a pilot and two test technicians on board. Fortunately nobody was hurt. Since there were buoyancy tanks on this early experimental ACV, the craft floated steadily upside down on the water surface, and all the crew crawled from the navigation cabin and up to the bottom of the craft, laid down and took a rest, and waited for help. This incident demonstrated that the ACV with not well-designed flexible skirt, in calm water and even in light stern wind, might experience dangerous flexible skirt tuck under, craft plough in, and even overturning in case of broaching. Since then, the design and operational guidelines have been established to prevent such dangerous accidents. This includes model testing in a ship test tank, which is done in China as it is in the UK, the USA, and Russia, and is an important part of the design process for hovercraft.

The gas turbine powered Chinese prototype AC landing craft type "722-II" was designed from the start with a responsive skirt [2-4]. The craft is shown in Fig. 2.10. Seakeeping improved significantly, and also the life of skirt, both bag and fingers; however, it can be a catch 22 situation. The author, as the chief designer of this craft, has such experience from sea trials. The craft, running at high speed down wind, ploughed in suddenly in a gust and caused some of the test crew to fall and slide along the floor in the main cabin from stern to bow (about 15–20 m), suffering injuries. The reason for the plough-in was the bow bag failing with a large horizontal split. Therefore, one has to design responsive skirt carefully. Bow and stern skirt bags are now designed with "rip stops"—vertical seams or sewn strips that prevent a long horizontal tear.

Skirt Developments

There were a series of plough-in accidents and craft overturns in the 1960s and 1970s as skirt and lower hull design was refined. Being the most numerous, SR.N5 and SR.N6 craft were most affected. SR.N5 craft had a very high performance, and commanders often allowed themselves to operate with high sideways drift speeds. Tuck-under of the skirt when drifting sideways down-wind across waves could lead to potential rollover as well as sideways plough-in. An SR.N6 experienced these kinds of conditions in extreme seas in 1971 and overturned while on a ferry service in the Solent between Portsmouth and Ryde. This accident resulted in a formal enquiry being set up by the UK Government, and to the UK Department of Industry sponsoring a technical evaluation to avoid a repeat event. The work lead to a document issued in 1980 called Stability and Control of Hovercraft—Notes for Commanders [2-5], that gave clear guidelines so that craft operations could be kept within a safe envelope of environment, speed and manoeuvre. This work had a fundamental effect on the hull lower surface geometries adopted since the investigation, introducing canted planing surfaces for both forward speed and sideways drift, and freeboard sufficient to prevent gunwale submergence and subsequent roll over.

Following the SRN6 overturn in the Solent and its investigation, the BHC designed their first craft with the revised safety features together with the operator

Fig. 2.10 (**a**) Front view 722. (**b**) 722 side view

Hovertravel—the AP1-88 passenger ferry. This changed from gas turbine power to diesels, riveted aluminium structure to lightweight welded marine aluminium, and ducted propellers; as well as the revised hull lower geometry. The AP1-88 has stood the test of time and has been built for a long client list. The design has recently been updated for larger passenger capacities up to 180 seats from the original 95 and re-designated the BH series by Griffon Hoverwork. The basic design has also been used for utility craft, particularly those for the Canadian Coastguard both in Vancouver and in the St Lawrence River.

Air Propulsion

Early ACV's used aircraft propellers for propulsion. The SR.N2 through SR.N4 and SR.N6 craft series all used open propellers developed using aircraft technology. While efficient at speed, they were very noisy; not very helpful for a ferry operating to terminals with people within a few metres. These propellers were not optimized for static or low speed operation. The first development step was to revise the blade design to match operation in airflow of zero to 70 knots. Later versions of the SR.N5 and 6, and the SR.N4 had purpose designed propellers. The noise did not reduce significantly as the tip speed had to be high so that disc loading could be realistic and so keep the propeller diameter within practical limits. On a craft the size of the SR.N4 it was not considered practical to install ducts around the propellers due to the additional structural weight. The SR.N4 craft (Fig. 2.11) therefore sounded rather like a Second World War bomber approaching closely, while the SR.N6 (Fig. 2.12) had more of a loud whine. During the 1970s, this sound was one that most inhabitants of South sea, at the southern end of Portsea Island, Portsmouth will remember well, especially the roar as the "N6" ascended Hovertravel's sharp ramp and then quickly cut power as the landing pad was so small (limited by the road just along the beach).

It was small racing craft in the UK that experimented first with ducted fans for propulsion, partly related to safety, and partly because industrial plastic bladed fans were inexpensive to buy. In the mid 1970s, several companies in the UK then applied this to larger 10 and 12 seat craft with considerable success, and this was repeated on the SR.N6 replacement ferry for Hovertravel called the AP1.88 (Fig. 2.13), which also moved to air-cooled diesel engine propulsion. Now, even the current world's largest ACV, the Russian Zubr, is propelled by a bank of three ducted propellers (Fig. 2.14).

The noise from ducted fans or propellers is much more acceptable to the public—still fairly loud close up, but not piercing in the way the N6 noise was.

Fig. 2.11 SR.N4 Mk 2 at Ramsgate

Fig. 2.12 SR.N6 at Southsea terminal

Fig. 2.13 AP1-88 at Southsea

The efficiency is way up as well, so that modern craft have a much better operating range and economy in service.

The Zubr's ducts are fixed, as on all modern craft. During the development period for the US Navy air cushion landing craft, Aerojet General designed a prototype with four swivelling ducted propellers, Fig. 2.15, and in Finland a utility craft called "Larus" Fig. 2.16, was built along the same design approach. Both these craft were technically successful, while the complex power transmission machinery meant that they were not commercial successes, remaining prototypes only.

Fig. 2.14 Zubr

Fig. 2.15 JEFF A and B

Fig. 2.16 Larus

Amphibious Hovercraft Development

The above paragraphs have centred on the sparks that started the fire in the UK in the 1960s and 1970s to develop a new type of marine craft. With the support of government funding and then some significant local commercial applications Saunders Roe, BHC, and then several other companies including Hovermarine, and Vosper Thornycroft were able to build a substantial industry for a couple of decades. What about other parts of the world, and what came before Cockerell's inventions, and why did they not create an enduring major new marine industry?

Part of the answer lies in the power-plant available. In Chap. 1 we described how fast marine craft have developed as the available power plants have opened up new possibilities. We talk more about this in the chapter on future prospects; for now though let us consider the development of proposals—patents etc.—from the mid nineteenth century onwards, that laid the foundations for Cockerell's air jet lift, as well as for sidewall hovercraft.

In the mid nineteenth century, at the same time steam power changed the design of ships, inventors made proposals for pumping air under the hull so as to reduce resistance. Proposals at the turn of the twentieth century addressed craft more like upturned dishes, for example the patents of F.W. Schroeder [2-6]. From the design sketches the inventors were clearly interested in the principle, rather than thinking about the potential payload and how to accommodate it. A boat was built by Sam Saunders in 1906 at the Saunders Syndicate yard on the River Thames, to test the theory [2-7]. Unfortunately, it did not prove as successful as expected. The air jets blown under the flat hull blew out bubbles, but did not propel the boat, so it was reported in the press at the time as the "bubble boat". Unfortunately, that was as far as the prototype was taken just then, Saunders attention being taken up with stepped monohulls.

Fig. 2.17 Kaario

We have to advance to the 1930s for the emergence of prototype craft that moved forward on an air cushion. Toivo Kaario in Finland was one of these pioneers. His craft, powered by an aircraft engine and propeller looked a little like today's air boats, except that the propeller is in front and some of the air is directed into a cavity under the curved wing shaped hull and the rest aft for propulsion (Fig. 2.17). These craft worked well over ice—useful in the winter in Finland, but did not result in development to commercial exploitation.

Later in the 1950s, in the USA, other pioneers also built "ground effect craft" following the same principle as Kaario, forcing air into a triangular shaped cavity under the hull structure to create an air cushion (Fig. 2.18). The possibility of

Fig. 2.18 Bertelsen Aircopter ground effect craft

scaling up this effect to warship size attracted the US Navy, and in the 1960s this resulted in a research programme to develop such a craft, with ducted fans at the bow to blow air under a space between two hulls, and separate propulsion engines aft. Today's wing in ground effect craft, WIG for short, came from an extension of this idea which we review in the next chapter.

The US Navy did not have much success developing their ground effect ships due to the prodigious power requirement so this programme died out, to be replaced by the surface effect ship (SES) prototype programme, see below, and the amphibious hovercraft programme targeted at a fast replacement for navy landing craft—the LCAC.

In China, the initial focus was on sidewall hovercraft, for use in the river system as passenger craft in the late 1950s and early 1960s. The success of Christopher Cockerell's designs encouraged research through the 1960s into amphibious craft at MARIC, the Marine Design Centre in Shanghai, leading to a number of prototypes. In China, marine engines built in country were not as efficient as those available in the West at that time, so until China opened up to the West in the 1980s, it was not possible to duplicate the kind of designs built by BHC. In the 1980s, technology had moved on in the West for amphibious craft, with welded aluminium hulls becoming the norm, the use of air-cooled diesels from Deutz for power rather than gas turbines, and the development of ducted propeller propulsion. MARIC were able to take advantage of this themselves in the 1980s and so introduce several amphibious craft for utility applications.

Fig. 2.19 Hoverferry terminal at Dover

During the 1960s, when the first commercial craft were introduced in Europe, the focus was on fast ferry applications. The amphibious properties were used at each end of the route where the craft would ascend a concrete ramp and land, rather than moor at a pier, Fig. 2.19. Saunders Roe, subsequently BHC was the industry leader, supplying craft of the SR.N6 series for short passenger routes (Fig. 2.12), and SR.N4 for the UK cross channel passenger and car ferry services (Fig. 2.11). Both these series were built with aircraft derived riveted aluminium hull structures, and used aircraft gas turbines (Rolls Royce Gnome and Proteus) driving aircraft derived open propellers. Through the 1970s and 1980s such passenger and car/passenger services were quite successful, particularly in the UK, Japan, and China. In other countries there was more success with sidewall hovercraft (e.g. in Hong Kong) before efficient and fast catamaran ferries won the hearts of many ferry operators.

In Japan, Mitsui were the first company to design and build ACV's in the late 1960s following the mix of aviation technology and skirt systems design pioneered by Saunders Roe. Mitsui's craft were aimed at ferry services where higher speeds could improve the service by increasing frequency. The routes their craft were used on were similar to the cross Solent route established by Hovertravel, with trip times of 15–20 min, for example at Oita airport across the Beppu bay to Oita city. Figure 2.20a, b shows the Mitsui MV-PP5, and MV-PP10. The MV-PP5 was a similar size to the BHC SR.N6 passenger craft at 16 m long and 8.6 m wide. Figure 2.22a shows the craft mechanical system was arranged a little differently from the SR.N6, with the IM 1,050 shp gas turbine mounted behind the lift fan at the back of the passenger cabin. Propulsion was from two variable pitch propellers, reducing noise emissions, and with the machinery all at the stern the internal noise level was also lower than the UK craft. The PP5 had space for 52 passengers and could reach 55 knots in its standard version and delivery services in normal sea states at 40 knots. Like the SR.N6, it was also built in a "stretched" length version at 18.2 m long with space for 75 passengers as shown in Fig. 2.20a.

Fig. 2.20 (**a, b**) *above* the MV-PP5 and *below* the MV-PP10

The MV-PP10, Fig. 2.20b was a larger craft able to accommodate 100 passengers. It was developed in the 1980s and adopted much of the same design approach as the AP1-88 in the UK, using four Deutz BF12L 513CP 440 kW air-cooled diesels for power, a battery of small centrifugal fans for lift located either side of the passenger cabin, and ducted propellers for propulsion. The MV-PP10 was a larger craft than the PP5 at 13.1 m long and 11 m wide. Top speed was similar to the PP5 at 54 knots, and service speed at 45 knots. The first craft was delivered to Oita Ferry company in 1990, and was followed with further craft in 1991 and 1995, finally in 2002 a

Fig. 2.21 SR.N4 Mk3

fourth craft was added just before the Football World Cup. The last craft had water cooled MTU 12V 183 TB2 440 kW diesels fitted for propulsion rather than the air-cooled Deutz. The craft remained in service at Oita until 2008.

Figure 2.21 shows the British ACV SR.N4 Mk3, the world's biggest passenger-car ferry hovercraft built and operated so far, on services across the channel between England and France. In its Mark 3 version, it accommodated 420 passengers and 55 cars at up to 70 knots in calm conditions often found in summer time on the route. The craft had a length of 56 m, had a beam of 25 m, and was powered by four 3,300 hp gas turbines, driving four lift fans (diameter 3.35 m), and four free air propellers as propulsion devices (diameter 6.4 m). The fleet of SR.N4 craft operated from Pegwell Bay and Dover Harbour across to Calais and Boulogne in France over a nearly 30-year period before being retired.

Two of the SR.N4 craft are now on show in Lee-on—Solent at the Hovercraft Museum in the UK (see resources at back of this book). The Hovercraft Museum is located at the former HMS Daedalus, a Naval Station that first housed the Interservice Hovercraft Trials Unit (IHTU) [2-8], and subsequently the Naval Hovercraft Trials Unit (NHTU). A short history of both these units is provided in Appendix 1. The trials carried out were important to amphibious hovercraft development in the UK throughout the 1960s–1980s. Many of these were in the form of military deployments where the ACV's performed a wide range of support duties in remote parts of the world in this period giving direct feedback that was used for design improvement.

Saunders Roe rapidly moved on from the SR.N1 experimental machine to the SR.N2 and SR.N3 aimed at passenger ferry and military patrol respectively [2-9] at the beginning of the 1960s SR.N4 got to the drawing board, but was rather ambitious as an immediate technology step and so the smaller SR.N5, a high powered

20 passenger craft was next out of the stocks in 1963. Initially it had air jet skirts, and it was only later in 1967–1968 that the change was made to an upper bag and lower "fingers". At the time Saunders Roe were developing the SR.N5 an agreement was made with Bell Aerosystems in the USA which resulted in Bell converting an initial batch of three craft for the US Navy and later another three were built from scratch in New Orleans by Bell for the US Army, designated SK-5 by Bell for service in the Vietnam war. The US Navy craft were designated PACV1–3 and the three Army craft, ACV901 to 903 [5-3].

The Navy craft (Fig. 2.22a) were unarmoured, and had simply a 0.30 calibre M60 machine gun and 40 m Grenade launchers as armament, and radar outfit to help locate the targets. The 'N5 was a fast craft being able to exceed 60 knots in calm water conditions and having a useful range of about 160 nautical miles. It was very noisy due to its aircraft propeller, so was no stealth craft. Approaching at that kind of speed though in wetlands became a fairly terrifying experience to the Vietnamese villagers and Viet Cong.

To enhance this, craft were painted with shark teeth around the bow skirt. The craft proved able to clear obstacles up to 1 m height, navigate ditches and canals, and to climb over sloping side rice paddy dikes up to 2 m high as well as pushing through grasses and sapling trees—basically not a lot could stop it advancing in the delta area.

The PACV craft were deployed first in 1966 and operated as experimental units attached to Navy Task Force 116. The craft proved very effective in cutting off known supply lines for the Vietcong as far as the Cambodia border along the Mekong and Bassac river delta waterway maze, as well as providing fast personnel logistics in these difficult areas, being able to insert and extract special forces personnel very quickly, and deliver them back to mother ships offshore if required, Fig. 2.22b showing a sortie from LHD USS Gunstan Hall.

Such lightly built craft would be easily damaged by enemy fire, but the PACV craft managed to avoid this by their extreme speed, and the noise, to which the Viet Cong responded by hiding. On night time patrols sampans were simply sunk by driving over them. The greater issue for these craft was their durability, both the structure and the skirts, on such demanding operations. The engineer support managed nevertheless to keep them running through 1966.

Their most famous operation was a search and destroy mission out of Moc Hoa in the "plain of reeds" south of Saigon in November 1966 that was very successful. The downside of the Navy craft was their lack of any protection for the delicate machinery, and so armouring panels were installed to protect the crew, and sensitive machinery and systems on the craft. The Navy craft were taken back to the USA for overhaul end 1966 and returned to service at end 1967 including new bag and segment skirts and higher power main engine. Their success resulted in the craft becoming a special target for the Vietcong during the later part of their operations.

Experience with the USN PACV lead the Army to modify their craft before deployment to Vietnam, lengthening the craft a little, and installing reinforced side decking and protective ballistic panels to their craft to make them more durable under fire (Fig. 2.22c).

Fig. 2.22 (**a**) Bell SK-5 PACV. (**b**) SK-5 leaving USS Gunstan Hall. (**c**) US Army PACV showing armour protection

Fig. 2.23 Griffon 475 RNLI

The army craft operated in the "plain of reeds" area from a base at Dong Tam. In parallel with this the Navy craft were moved north to Tan My close to the town of Hue. The more heavily protected army craft continued the reconnaissance, personnel insertion and direct assault missions, often working two craft together with a helicopter. The three craft ACV's 901, 902 and 903 were deployed in May 1968. ACV901 and 902 were destroyed in action in January and August 1970, and only ACV903 has survived to be returned to the USA and is on display at the Army transport Museum, Fort Eustis, VA. In their short life in service, the craft showed the high work capacity of ACV's, and also the very high maintenance demand from an aerospace craft operating in a marine environment. Skirts and gas turbine maintenance were a significant part of the demand as well as battle damage. Both the Navy and Army concluded that these deployments could not be sustained and so operations were discontinued after 1970.

The learning from the US Army and Navy use of ACV's in Vietnam was strongly positive from the point of view of the concept potential. On the other hand the aerospace technology was a strong handicap. The British joint forces gathered similar experience operating SR.N5, N6, and a number of other designs on trials, exercises, and paramilitary operations. After 2 decades of operations the British establishment was closed (see Appendix 1) but contributed much to the next generation of craft, both commercial and military. Since the early 1980s amphibious ACV applications have focused on special utility and paramilitary operations, such as craft for coastal patrol, and for rescue in areas that boats have difficulty.

These duties have developed a market for craft with a payload as small as 6 people, for example the Griffon 475 used by the British RNLI, Fig. 2.23, and up to

Fig. 2.24 Zubr in operations

very large craft for offshore patrol such as the Zubr, designed and built in Russia, Fig. 2.24. Zubr is the biggest military ACV currently in operation, weighing 550 t, achieving 60 knots speed in calm water, accommodating three tanks of medium size, or one tank plus 600 marine troops. It is a coastal patrol ACV rather than a craft operating from a mother ship deployed to far sides of the world. The craft is now in operation in Greece as well as in Russia, enabling fast deployments of troops and equipment to remote parts of these countries' territory.

The oil industry and also the mining industry has been a regular customer in the last 20 years for personnel transfer to drilling sites, and also for logistical load deliveries. Heavy lift craft Larus (Fig. 2.16) built by Wartsila in Finland was intended for these kind of operations and carried out extensive test operations in Finland in the early 1980s and Canada from 1986, leading to better understanding of stresses on diesel engine operations in low temperatures, skirt durability, and manoeuvring performance in the ice climate, which is much rougher than one might imagine due to continual break-up and rubble formation in coastal areas. The demand is now resurging for craft suitable for arctic and shallow water operations, as climate change opens up the Arctic ocean rim to the oil industry. In the 1980s, skirt design and rubber impregnated fabric technology was still in an early phase. Endurance was limited and lead to significant maintenance demand. This is now being improved through current R&D programmes for new planned operations [2-10].

On the military side, apart from amphibious coastal patrol, the major amphibious ACV development over the last 30 years has been the US LCAC programme. Over 100 of these craft have been built and are operated out of two bases, one each on the Atlantic and Pacific coasts of the USA. The craft operate from mother ships (US Navy LHD class) and allow delivery of troops and equipment ashore from over the

Fig. 2.25 (**a**) LCAC under way. (**b**) LCAC inside LSD loading up

horizon at high speed. The LCAC gives access to 70% of seashore worldwide, rather than 17% for conventional landing craft, Figs. 2.25 and 2.26. In Russia, a parallel programme for ACV landing craft was carried out, with the craft mothered to amphibious ships mainly stationed in the Pacific coast of Far East Russia, their craft is called the "Lebed" (Fig. 2.27).

Fig. 2.26 LCAC coming up on beach

Fig. 2.27 Lebed

The LCAC programme has been through design improvement cycles more than once since the initial build programme, with a major step forward being the second generation deep skirt system that was introduced early in the 2000s giving higher performance and reduced spray generation. A further redesign of the power and mechanical transmission system has been undertaken for implementation in

Fig. 2.28 BHT130 Solent express between Ryde and Southsea

replacement craft, as the original batches reach their design life and have to be retired after their 30-year economic lifespan. The existing LCAC craft statistics are all up weight 150 t, payload 50 t (one medium size tank), speed 50 knots, installed power 2×10,000 shp. Since 2007 the US Office of Naval Research has been carrying out studies and engineering for a longer term replacement for the LCAC called the Transformable craft. This is rather larger than the LCAC, aimed at transfer of payloads between 300 and 500 t at a time delivered from much farther offshore. The T Craft uses elements of the Catamaran, SES and ACV so we discuss this in detail in Chap. 8, meanwhile the LCAC continues to deliver a high capability to the US Navy and Marines.

The following series of figures illustrate various types of ACV in operation for different applications, to give an idea of the wide range of utility duties these craft now undertake:

1. *Short range Passenger Ferry*: Figure 2.28 above shows the BH 130 now in service across the Solent with Hovertravel.
2. *Coastguard ice breaking and river navigation marker maintenance*: Figure 2.29 shows the Canadian Coastguard BHT-150 Mamalosa on the St. Lawrence River near Montreal. The craft is used for ice break-up maintenance on St Lawrence tributaries each spring for flood control, and for navigation aid maintenance throughout the year.
3. *Recreation*: Figures 2.30 and 2.31 show an early craft running on a beach and coming up a steep incline after a river run. Hovercraft use momentum as well as thrust to make such ascents. It can be exciting, and also a challenge in tight situations at the top of the ramp so swivelling for reverse thrust close to the top and killing lift has be used for precise positioning.

Fig. 2.29 CCG BH150 Mamalosa over broken ice on St Lawrence near Montreal

Fig. 2.30 Recreation Hovercraft at the beach

4. *Remote Village supply*: Figure 2.32 shows a half deck AP1-88 in service in the Aleutian Islands in Alaska that transports supplies and vehicles to remote villages and outposts, coping with the winter ice, and beaches for terminals.
5. *Coast guard*: Figure 2.33 shows a Swedish Coastguard Griffon 2000TDX on winter service in the middle of an ice field in the Western Baltic.

Fig. 2.31 Recreation craft taking a steep ramp from a river

6. *Exploration and humanitarian aid*: Figure 2.34 shows a River Rover craft of the Missionary Aviation Fellowship sunning through a severe grade rapid on the river… in Peru.
7. *Frontier Patrol*: Figure 2.35 shows a Griffon 8000TD running ashore across shallow waters at the coast of Saudi Arabia for the Saudi Frontier Force. Similar craft are in service in Pakistan and India in areas where waterways are very shallow.
8. *Operations over marshland*: Figure 2.36 shows a Griffon 375 speeding over a marsh. An excitement for the passengers, particularly over bumps and gulleys, an operation unique to the ACV.
9. *Smuggler Patrol*: Figure 2.37 shows a Griffon 8000TD of the Indian coastguard approaching a boat suspected of illegal activity.
10. *Pollution control*: Figure 2.38 shows oil containment booms deployed to a beach for laying out across shallows to protect against oil slick damage and contain for removal.
11. *Civil Engineering Support*: Figure 2.39, shows a craft supporting a crew performing survey and taking soil samples at an intertidal location in Pakistan in preparation for piling and structure installation.
12. *Hydrographical Survey*: Figure 2.40 shows a Griffon 1000TD making a hydrographical survey in a tidal area and taking soil samples through a small "moon pool" in the craft bottom.
13. *Ice Rescue*: Figure 2.41 shows a casualty being rescued from broken ice. There are a number of craft operated by Fire Services in the UK, the USA, and Asia as rescue vehicles. These are generally small—four to six persons payload and

Fig. 2.32 (**a**, **b**) AP1-88 Village supply, Aleutian Islands

are able to reach the scene quickly and safely if someone falls through a patch of thin ice.

14. *Seismic survey*: Figure 2.42 Shows a craft deploying an array of seismic charges and transponders that will provide a detailed picture of the deeper subsurface below this tidal flat and perhaps identify the local aquifer, or deeper down,

Fig. 2.33 Griffon 2000TDX Swedish coastguard craft

Fig. 2.34 Hoveraid river rover craft in Nepal on river rapids

hydrocarbons. Rapid deployment to site is essential for such work at sites like this, as setting the charges and transponders is a significant task, and the completing the shots, as well as equipment recovery all has to be done before the tide returns!

Fig. 2.35 G8000TD craft on shallow waters

Fig. 2.36 G375 over marshland

The future for amphibious ACVs looks likely to continue to be for these special utility operations, designed with ducted propeller air propulsion, welded marine aluminium or fibre reinforced resin hull structure, and flexible skirts that will continue to develop as operating experience guides the demands for spray control, damped surface response and low wear rates even in subzero temperatures.

Fig. 2.37 G8000TD on smuggler patrol in India

Fig. 2.38 Craft performing pollution control deploying floating containment boom

The ACV for utility and commercial service has matured into a size range dictated by the available high-speed diesel engines. Skirt technology has advanced with improved understanding of its interaction with the hull and excitation from waves as a lightly damped spring system. Now, it is possible to design a loop and segment skirt that responds to the sea waves and provides a relatively smooth ride while also having acceptable wear rate of the segment tips for commercial operations. The current challenge is spray control. This is a key issue for the LCAC, and

Fig. 2.39 Pakistan civil engineering support

Fig. 2.40 (a) Hydrographical survey. (b) Taking core samples via moon pool

Fig. 2.41 Craft performing ice rescue

Fig. 2.42 G2000TD seismic survey on tidal sandbank, preparing a shot

Fig. 2.43 LCAC over land showing secondary segments for spray control

also for craft operating in ice/snow conditions. Craft such as the Solent ACV ferries use an "apron" draped over the loop around the bow and bow quarters (Fig. 2.28). LCAC and arctic craft need a more sophisticated system based on secondary segments. Designs for this are emerging at the current time (Fig. 2.43).

Propulsion systems have changed dramatically over the years. The breakthrough in this area was the adoption of a duct around the propeller, and a redesign of the impeller along the lines of an axial fan to reduce blade loading and noise emission. The induced flow around the duct increased thrust at low speed, and allowed the propeller diameter to be reduced. Noise was significantly reduced, and with careful design the duct also could be used to mask some of the emission. The duct also assisted directional control as it formed a "fin" bringing the craft centre of vertical area aft of the centre of gravity. While larger craft use variable pitch propellers, smaller craft such as the AP1-88 use rotatable air nozzles in the bow fed from cushion fan air to manoeuvre the craft at a terminal. Ducted propulsion systems are now in use on the LCAC, and also the Zubr.

In the 1990s many utility craft used the Deutz air-cooled diesel, as this was an efficient unit and was very simple to install. More recently as demand has grown for larger size craft, builders such as Griffon (now merged with Hoverwork who built the AP1-88 series now developed to the BH series) have ventured into liquid cooled diesel power. The power units come from large trucks and are relatively light weight. The market for road vehicles is very large and with the drive to greater efficiency and lower emissions, we have seen the emergence of diesels with turbo-charging, direct injection, and adaptation to biofuels or a mix. These developments will continue, and so ACV's should also be able to reap the benefit of the efficiency advances.

The main addition that has to be made for hovercraft is marinisation—protection of the materials against salt water and spray. This is most important for radiator and other cooling system components, and for design of the air feed to the motor itself. Design of air intake filter systems had advanced also over the last decade or so, and so long as maintenance (washing/cleaning) is diligent, operations will remain reliable. The Canadian, Swedish, and Finnish Coastguard, working in sub-zero temperatures in winter, have a demanding maintenance regime, taking care of snow and ice accretion on the craft and filters. This has required incorporation of heating in certain parts of the craft, and for the diesel fuel, similar to the outfit on trucks operating in the Northern parts of Sweden and Canada.

There is still much potential for improvement in ACV design, targeted mainly at controllability and operational endurance rather than pushing operational speeds higher. Operation in the 30–50 knots speed envelope seems to fit with the market, through a balance of relatively short deployment time, while maintaining economy.

Sidewall Hovercraft

At the time Christopher Cockerell developed the first peripheral jet hovercraft a study team in Harbin Military Engineering Academy, China, was starting research at the end of 1957 on development of an air cushion craft with an air plenum chamber.

The plenum chamber lifting principle is shown in Fig. 2.44, in which, the lift fan one, driven by lift engine two, blows pressured air into the air cushion plenum

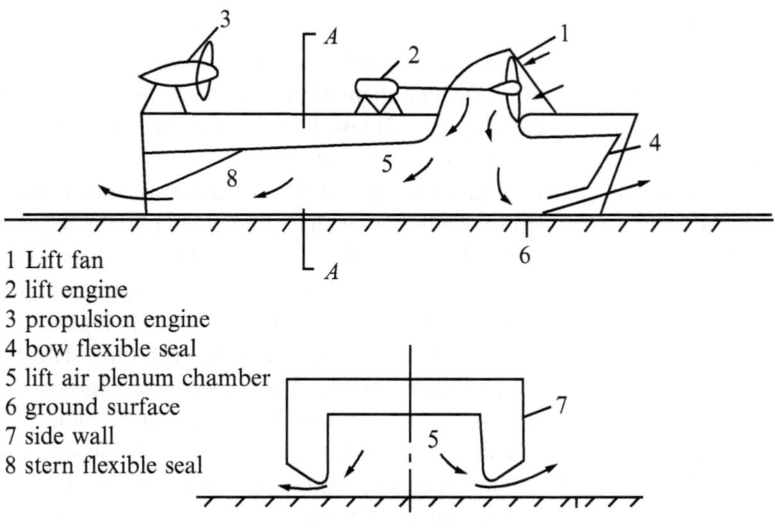

Fig. 2.44 Sidewall SES plenum chamber principle

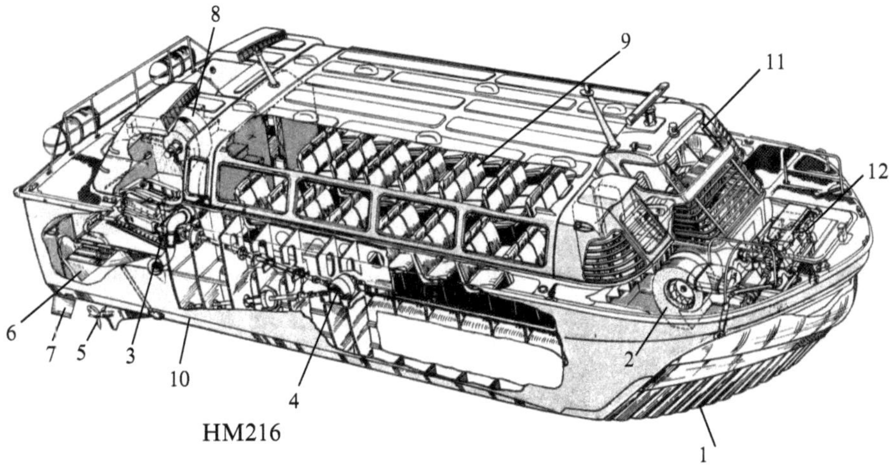

Fig. 2.45 Hovermarine HM2 Cutaway

chamber and lifts the craft from ground or water. The air cushion is sealed on each side by a sidewall, seven, and at the bow and stern with movable seals four and eight. The prototype craft was propelled by an air propeller three at the stern.

The prototype sidewall craft, type "33" was designed to be amphibious to a limited extent even though using solid side seals. This is shown in Fig. 2.9 where (a) shows the craft on land, and (b) over water. The craft had an all up weight (AUW) of 4.3 t, driven by aviation piston engines of 176.4 kW for lift and 117.6 kW for propulsion. It carried out its first sea trials off the port of Lu Shun, achieving a speed of 69.5 kph during a trip of 16 nautical miles. The hovering height, i.e. the clearance between the bottom of sidewall and ground was very small, 10–20 mm, and the craft could only be operated and landed on smooth sand beach. It did not really possess practical amphibious capability. Development of the craft therefore turned to developing its purely marine capability as a sidewall hovercraft, or SES with flexible skirts added to seal the bow and stern of the craft.

The physical concept of air plenum type ACV and SES was almost the same; however, the latter is designed so that the sidewall is submerged into the water during the craft operation at service speed, saving a large amount of lift air flow, so that the lift power of a sidewall hovercraft is only 1/3–1/4 of that of an amphibious ACV.

Examples of early SES designed in the 1960s–1980s are shown in Figs. 2.45–2.49. Figure 2.45 shows a cutaway view of British sidewall hovercraft type "HM216", in which 1-bow skirt; 2-lift fan; 3-propulsion engine; 4-V transmission gearbox; 5-water propeller; 6-stern skirt; 7-canted rudder; 8-stern lift fan; 9-passenger cabin; 10-sidewall; 11-pilot and navigation cabin; 12-bow lift engine; Fig. 2.46 shows a Chinese passenger SES type "717c" with water jet propulsion, running in rapids; Fig. 2.47 shows the bow skirt of Chinese passenger SES type "719II"; Fig. 2.48 a Chinese SES Ferry operating between Hong Kong and mainland China; Fig. 2.49 a high-speed passenger ferry in South Korea, the Democracy 1.

Fig. 2.46 SES 717c

Fig. 2.47 Bow skirt of SES 719 II

Fig. 2.48 Chinese SES ferry en route Hong Kong to Guangzhou

Fig. 2.49 SES passenger ferry democracy 1 in Korea

SES differences to amphibious craft are:

- By saving lift power the high-speed marine diesel could be used as the main engine to replace the aviation engines used on early ACV's, saving on both procurement and maintenance costs.
- Higher propulsion efficiency by using marine propellers rather than airscrew propulsion.
- Lower drag at low speed due to reduced water friction drag on sidewalls compared to the drag on skirt segments up through the range to main drag hump. The drag peak itself is also lower due to using a higher L/B.
- Higher drag at high speed due to immersion of the sidewalls above hump speed.
- Lower transverse stability for thin sidewall craft such as the Hovermarine HM2 and HM5 series. This can be corrected with wider planing side-hulls, though with the consequence of higher drag forces.
- Ability to scale up to larger size than the amphibious ACV, particularly for craft using wider side hulls.
- Large deck area for accommodating low density payloads such as passengers, cars and Ro-Ro trucks and trailers.

Chinese SES Experiences

Risks on the Upstream Stretch of the Yang River

At the beginning of 1978, in order to evaluate the possibility of using sidewall hovercraft (SES), operating in river rapids and shallow water, the SES type 717c test craft (Fig. 2.46) was taken to operate in one of the upper tributaries of the Yang Zi River, called Jin San River. The leading particulars of 717c are: $L \times B \times H$ (m) = $20.4 \times 4.5 \times 5.4$, two diesels driving two water-jets as propulsion, service speed 45 km/h with passengers and test personnel up to 40 aboard. The craft was considered suitable for test operation in the Jin San River as it was designed with water-jet propulsion. This stretch of river is a most dangerous route, especially at the "Shu Shoal" (also called the Ghost Shoal). Many small boats and junks have had accidents here from generation to generation. No commercial navigation was undertaken there at that time for this reason. However, it is a very important river, since the biggest city, at the west part of Sichuan Province, Yi Bin City is located here, at the intersect point of both Yang River and Jin San River. Downstream from the city was Yang Zi River, and upstream from this city is Jin San River. A lot of important national resources such as iron mines, hydropower generators, etc., are located along the Jin San River.

However, at that time the Jin San River still had not been exploited, so the government asked MARIC to investigate the possibility of developing a route there along the river for transportation of important materials and people. A test team of 13 persons including three boat crew had been organized, and started the trial voyage of SES type 717c at the end of 1977 from "Yi-Bin City" to "New Township", a small town 150 km from the Yi Bin City along the Jin San River.

The Jin San River has shallow water rapids, and an inclined river bed, as well as a lot of shoals along the river. Around the river are a lot of high mountains. In the early morning, we started from Yi Bin and transited the rapids smoothly, with a very happy mood, since we were heading upstream with nice handling and manoeuvrability using the water-jet propulsion, and the landscape was so beautiful there, viewed from the deck of a fast craft between the high mountains with eagles seen above in the air. At the noon we arrived at the "New Township" by the mountain. This was a town where a lot of ethnic minority people lived, who hosted us with very friendly and kindly mood, and the governor of the town invited us to take part in a banquet, and enjoyed the beautiful dance performed by a troop of local girls.

After dinner, we were running back to Yi Bin City. This was a rough time, since the craft was running on rapids following the downwards flow and had poor handling. When we reached "Shu Shoal", the most dangerous shoal on the Jin San River, I could clearly see the captain's excited and nervous attitude, since he was a senior crew working in the Yi-Bin Water Transportation Company for a long time, whom we invited as the captain of the SES, and he knew very well that a lot of sailors died here previously, and nobody could escape from this rapids; even a champion swimming sportsman of the province died here some years ago. The river at Shu Shoal is a complex bend though a rocky gorge with very deep holes in the river bed under the water surface with depth much as 50 m compared with the ordinary river depth here of 2.5 m, causing very powerful rips and vortices. The flow was up to 7 m/s, i.e. 25 kph through the shoal.

Since at the left side of the craft were the shoal and large vortex, and at right side was a cliff, the SES had to run to the right side and close to the cliff, in order to avoid getting into the vortex and shoal; however, unfortunately, the craft lost handling due to the rapid stream and the effect of vortex, and rushed to the big rock, and landed on the stone. The craft was inclined, with right sidewall landed on the rock, left sidewall in the water with only 3 cm freeboard, with a hole on the sidewall, and strong rapids of 21 km/h under the deck. The situation was extremely dangerous, and nobody from the crew would be alive if the craft had overturned and people dropped down into the water.

Fig. 2.50 HM218 Tacoma port authority fireboat

As the technical director of the ACV division at MARIC, and being a leader of this test team, the author ordered all of members to stay calm, and not move so as to avoid destabilizing and overturning the craft, since the only possible method to escape from the accident was waiting for salvage. Fortunately, all of the team members were brave and nobody left the craft to land. After a dozen minutes, an engineering tug boat sailing from Yunan Province (Jin San River was a border river between the Sichuan and Yunan Provinces), arrived at the SES, so we asked them to tow the craft out from the rock, even though it was risky.

Fortunately, they successfully pulled the SES off the rock, and it could be lifted up on air cushion to drain out some water. The craft was hovering rather heavily listing to one side, but due to the sidewall filled with solid foam buoyancy as a safety measure during design there was adequate reserve buoyancy. The SES was still able to hover on its cushion, and completed the journey arriving at Yi Bin that evening. The achievement of surviving such a dangerous event was attributed to the air cushion technology. After this event, we received a lot of ACC orders from Sichuan Province!

European and USA SES Development

SES development in Europe and the USA began with the Hovermarine HM series of ferries (Fig. 2.45), utility variants of the HM-2 for harbour fire services (Fig. 2.50), and in the USA the push towards the "100 knot Navy" by Admiral Zumwalt.

The strategic idea for the SES as part of the bigger plan was originally to make a step towards mini aircraft carriers, based on the approach that if the SES could

Fig. 2.51 SES 10,000 mini aircraft carrier

run into wind at 80 knots aircraft could take off and land with a much shorter run. The much smaller vessel would allow more to be built and spread the tactical advantage and damage risk. This was considered particularly important during a potential modern naval war. Figure 2.51 shows an artist impression of an SES mini Aircraft Carrier. The projected SES100, and SES3000 would lead to this goal of the SES10000.

This programme began with a design competition, and building two prototype SES the SES 100A built by Aerojet General (Fig. 2.52), and the SES 100B by Bell Aerosystems, Fig. 2.53, who also built BHC SK-5 craft under licence for service in Vietnam in the Mekong delta during the late 1960s. The two craft were selected to be different, with the SES 100A having water jet propulsion while SES100B had surface piercing propellers. The seal systems were also different at the bow, with the SES100A having a swept back geometry, and SES100B having the bulbous responsive skirt geometry.

Both craft achieved their design objectives, while SES100B was a little faster in calm water. Eventually a third company Rohr Marine was selected to develop the full scale 3,000 t SES, and worked out the preliminary design. Unfortunately, the programme was cut at the time of the Middle East oil embargo in 1974, and the US Navy turned to SES at smaller size for mine countermeasures and coastguard use; and to development of the LCAC. The SES100 programme nevertheless drove forward the technology very quickly in just a few years. Figure 2.54 shows a guided missile successfully launched on US SES-100B at speed of 60 knots.

Reverting to rather slower speeds, Bell Halter, successor to Bell Aerosystems, designed the aluminium hulled BH-110, a diesel powered SES with cruising speed of 35 knots. Bell Halter was the first in the West to go for sidewalls that were broader in their centre section so as to provide space for the engines lower in the hulls. This configuration had advantages for stability but at a penalty of higher drag forces.

Fig. 2.52 SES 100A

Fig. 2.53 SES 100B

Fig. 2.54 SES 100B missile launch

The hydrodynamic flow around the stern section was complex, and influenced by the air cushion, so that air release under the keel in higher sea-states provided a challenge for the propeller design. The craft was configured as a crew and cargo boat aimed at the offshore service industry in the Gulf of Mexico. While this market did not take off—the crew boat operators being a conservative bunch, and happy with the speed of their crew boats already—the US Coastguard did take interest and eventually bought three craft for use on coastal patrol off Florida (Fig. 2.55). These craft were very effective in their duties, but operating costs were a problem even for these craft. The original BH110 was modified by the US Navy by lengthening to become the SES-200, Fig. 2.56. This craft was used as a demonstrator making visits to Europe and the Mediterranean, and providing a spark for the Norwegian military SES development. USN interest moved across to catamarans for their tactical support craft, which the SES-200 was originally meant as a prototype.

Fig. 2.55 SES BH 110 coastguard craft

Fig. 2.56 SES-200

There were many lessons from this sequence of developments in the USA. The compromises required to match broader side hulls to the propulsion system was one aspect, while the benefit of using simple 2D bow and stern seals for improved maintenance costs was another. Moving to welded marine aluminium for the hull was a success for the BH110 and demonstrated that the increased hull weight was not a disadvantage for medium speed craft. The cushion of an SES essentially adds another dimension to design of a catamaran. Cushion dynamics produced challenges on craft that were designed for higher speeds such as the SES100 prototypes, and lead to damping systems being installed—basically air vent valves that were tuned so as to dampen internal pressure pulsations that would start the craft vibrating up and down on its cushion due to the elasticity of the trapped air in the cushion bubble.

The Aerojet SES100A used water jets with a forward facing intake under the sidewalls. This system worked effectively, though as water jet design for fast catamarans developed in the 1980s and 1990s, the flush inlet water jet system demonstrated significant efficiency advantages. These can be applied to SES so long as a suitable fence is installed at the sidewall inner wall so as to prevent cushion air being entrained.

Another lesson relates to the operator capability and expectations. The SES 100 craft had to be built with sophisticated structures more aligned with aircraft maintenance practice than marine practice. Skirt systems were a wholly new technology. The power requirements meant that such craft did not have a long endurance. If the SES3000 had been put into service it would have required a major investment in new support facilities, as well as new operating practices in the US Navy. The BH110 was much closer to "normal" marine practice and thus lower technical barriers to its introduction. The low endurance and independent operational range meant such craft had to operate from close support bases rather than as independent strategic elements. This is OK for coastal patrol and mine countermeasures, but not for intercontinental power projection as practiced by the USA and NATO Countries.

Norway

In Norway, commercial SES production was started by Brødrene Aa in the mid 1980s with the SES Norcat using wider side hulls, high-speed marine propellers driven by diesels, and construction in FRP/sandwich to achieve light weight.

The design achieved higher speeds—close to 50 knots—than the catamarans in service along the Norwegian coast at that time, and the prototype was put into service between Tromsø and Harstad. Several more of these craft were produced (Fig. 2.57), financed by ship operators in Oslo and leased to Ferry Companies in the Mediterranean, West Africa and Caribbean. Some operations did not last more than a season, while others were more enduring.

The prototype initially had a problem with cobble-stoning, giving an uncomfortable ride in short sea states, but this was resolved through the installation of a "ride control system" which allowed small amounts of air to vent in a controlled manner providing a damping mechanism. Maintenance of the seal system was, and remains a specialist SES task, which adds to operating costs compared to a catamaran, so

Fig. 2.57 Brødrene Aa SES moored by the construction yard

once catamaran performance came closer to the SES, the market died away in the 1990s. Many of the fleet of SES constructed by Brødrene Aa are nevertheless still in service, having passed from operator to operator as they established themselves in competition with the traditional ferry companies and modernized their fleets, demonstrating their basic economy and reliability (Fig. 2.58).

The succeeding development step was the military SES, specifically the MCMH. In the early 1990s the Royal Norwegian Navy needed to replace its fleet of ageing displacement mine hunters and mine counter measures vessels. The SES offered an opportunity to improve safety—it had been proven through trials in both the UK and the USA that ACV's were quite immune to mines exploding close to them—and at the same time improve reaction time to this type of threat.

A design was developed, leveraging on the expertise developed through the Norcat experience and put out to tender for detailed design and construction. Kvaerner Båtservice Mandal won the contract and built the series of Kvitsøy class MCMH craft. The MCMH is propelled by flush inlet water jets rather than propellers, as these have a lower noise signature, as well as now being more efficient following a 20 years of active R&D for fast ferries of all types. Figure 2.59 shows two of the craft in Stavanger harbour during summer 2009. The Kvitsøy craft are quite active along the Norwegian coast and are notable as they are rather quicker than other vessels, seen from the coastline.

Fig. 2.58 Brædrene Aa CIRR-120P "Wight King" at Southampton

Fig. 2.59 SES mine hunters in Stavanger 2009

The latest programme, in Norway is for a class of fast patrol craft, the Skjold class, developed jointly by UMOE Mandal Shipyard and Kongsberg Defence and Aerospace. This is gas turbine powered and can reach 60 knots in coastal conditions at full power, Fig. 2.60. The latest flush inlet water jet technology is incorporated,

Fig. 2.60 (a) SES Skjold fast attack craft. (b) Skjold passes an iceberg at speed

with planing side-hulls, and active ride control systems. The first of class was tested in 2007 and following a lengthy proving programme the basic vessel was accepted by the Navy in 2009. Since that time further work has continued with the combat system development and installation. The first craft for active service was handed over to the Royal Norwegian Navy as FPB P961 "Storm" on 9 September 2010. The second of class FPB P962 "Skudd" was accepted on 28 October 2010. The remaining four vessels will be delivered through 2011.

Japan

In Japan the early focus was on amphibious passenger ACV at Mitsui with the MV-PP5 and MV-PP15, but in the late 1980s work began on a ship sized SES for commercial use as part of a competition between the SES and Hydrofoil concepts; the Techno Super Liner programme from 1989 to 1995.

The programme began with an 18.5 m test craft built by Mitsui at its Shimonoseki shipyard, the Meguro. The craft was initially powered by four IHI IM-100 gas turbines, tow of lift driving banks of centrifugal fans and two for propulsion driving water jets. Following initial trials in 1991 the craft was lengthened to 25 m and propulsion power uprated with two Allison 501K gas turbines rated at 3,185 kW. The lift turbines were rated at 772 kW. In this latter form the Meguro two achieved speeds exceeding 60 knots and gave much encouragement to the development programme for SES which continued with the 70 m prototype.

Figure 2.61a shows a cutaway of the 70 m length, 3,000 t displacement prototype TSL-A craft showing its power and propulsion layout. The items in the diagram are 1-bow seal and cushion lift fans, 2-bow cushion lift engine, 3-Bow skirt seals, 4-Ride control vent ducts, 5-bow hydrofoil stabiliser, 6-hull structure, 7-starboard main propulsion gas turbine, 8-planetary reduction gearbox, 9-water jets propulsors with scoop inlet, 10-rear cushion multi-lobe seal, 11-rear cushion lift fan, 12-rear cushion engine. Figure 2.61b shows the prototype TSL-A "Hisho" in operation with containers on board. The initial mission target for TSL-A was high-speed coastal cargo service using marine containers. Figure 2.62 shows the SES concept developed, the TSL127 with a full load of containers. The TSL design had a high L/B for reduced drag at high speed and also reduced hump drag; flush inlet water jet propulsion; and wide side hulls for stability. The cushion cavity is deep to offer high quality seakeeping offshore. The vessel was completed in 1994 and put through extensive sea trials including operation for about 4 months of the typhoon season. During the trials the craft travelled 17,000 nautical miles at speeds between 40 and 50 knots in various sea states to verify seakeeping and performance in both day and night operations. Cargo delivery tests were also carried out using normal 20 ft marine containers taking fresh food, flowers and industrial products. It was found that all the products were safely delivered without any degradation due to their high-speed delivery! This was an important result for possible future commercial operation of the high-speed craft.

TSL-A Hisho was converted for utility service and put into operation as a disaster salvage craft at Kobe Port and renamed "Kibo" (hope in Japanese). After the first 2 years of operation the lease to the city was not renewed as the fuel and operating costs were high, and the service able to be provided was not economic. The craft was converted to a car and passenger ferry in 1997 and put into seasonal service as a ferry between Shimizu and Shimoda in Shizuoka Prefecture east of the port of Kobe, Fig. 2.61c. In this form it went into service for a while. At the beginning of March 2000 before the main summer season the Kibo made a trial run from Nagasaki to Shanghai, covering the 740 km distance in close to 11 h before being inspected

Fig. 2.61 (**a**) TSL cutaway drawing showing systems outfit. (**b**) TSL A Prototype "Hisho". (**c**) TSL-A reconfigured "Kibo"

Fig. 2.62 TSL 127 artists impression

by Chinese authorities. Whilst proving the route technically, the trial did not result in a service. Fuel costs also became a problem for this subsidised service also, and the craft has now been decommissioned from regular service. It may still be able to be seen in Shimizu Port.

The final stage of the TSL-A programme was construction of the 140 m length "Ogasawara", a vessel of 14,500 t displacement built by Mitsui in 2003–2005 as part of the millennium projects of the Japanese Government. The vessel was built as a fast ferry to connect Tokyo and Chichi Jima in the Ogasawara islands to cut the journey time from 24 h and more to about 18 h by travelling at approximately 40 knots. The go-ahead was based on the success of Kibo operating as a ferry. The vessel budget was 11.5 billion yen. It was completed as far as sea trials, achieving 42.8 knots in 2 m seas in October 2005, Fig. 2.63, but was stopped before final handover to the operator, as the shipping company calculated that the service would lose at least 2 billion yen each year, and the government could not commit to subsidise this. Details of the craft are given in Table 2.2. A sad ending to a solid technical programme, and construction achievements not yet reached outside Japan.

It is instructive that TSL-A is similar size though with much lower service speed when compared to the SES 3000 proposed for the US Navy in 1974. The economics driven by fuel cost for such large high-speed vessels have not fundamentally changed over the ensuing period. Comparing with the Norwegian SES programme it is useful to note that lightweight fibre reinforced resin is used for the hull construction of the smaller Norwegian vessels keeping the displacement as light as practicable. Some catamaran builders have also followed this approach, Brødrene Aa now build carbon fibre and resin hulled fast catamarans instead of the FRP SES that they built through the 1980s.

Fig. 2.63 SES "Ogasawara"

Russia

In Russia, there has been a long history of air lubricated craft design for craft operating along the Volga—Don river system, starting in the 1960s. These developments paralleled development of hydrofoils, where simpler craft were needed on shorter trips for passenger ferries. Examples are listed in Table 2.2, with some statistics. Like Russia's river hydrofoils, significant numbers were built and operated successfully. This design series was then expanded to design of sidewall hovercraft, with bow and stern seals to give a deeper air cushion and smoother ride. The learnings in developing these craft has been extended to the world's largest military SES built to date, the Bora class SES corvettes, of which there are two in service in the Russian Black Sea fleet, the Bora and the Samum.

Figure 2.64 shows Bora, at anchor, and leaving harbour. The ship is 1,000 t displacement, length 65 m, and 55 knots dash speed with range 800 nautical miles and 12 knots cruising with range 2,500 nautical miles. The vessels have has been operational in the Russian navy since 1989. Type SS-N-22 guided missiles are an important part of the ship's armament.

Sweden

Sweden has also built and tested a "stealth" fast strike craft SES at the end of the 1990s, Fig. 2.65 shows the craft, called "Smyge". This was also constructed in lightweight carbon fibre reinforced resin.

Fig. 2.64 SES Bora patrol frigate based in Russian Black Sea fleet. (**a**) SES Frigate Samum returning to port. (**b**) SES Frigate Bora leaving port

China

The SES continues to enjoy some success as a passenger craft in China on the Yang Zi river, with several craft marques that are built by shipyards in China voyaging much of the lower river between the main towns down to Shanghai. SES ferries are

Fig. 2.65 Swedish fast attack stealth SES, Smyge

also in service between Hong Kong and Guang Zhou on the Pearl River. As outlined above, on a worldwide basis the trend has been towards catamarans for ferry service and SES development for special naval applications in the last decade or so, with longer L/B, and more sophisticated cushion systems. The largest SES depend on gas turbine power, just as monohull naval frigates and destroyers do in the modern Navy.

Further Development

What characterizes ACV and SES performance, and the challenges to its further development? Let us consider the main issues of cost, performance in a seaway, physical dimensions and limitations in size.

Cost, Fuel Consumption, and Technology

ACV and SES are complicated craft, incorporating sophisticated structures, cushion seal systems that require regular attention to remain efficient; high installed power, often with gas turbines as the prime movers; and sophisticated propulsion systems. Their capability is high; particularly for commercial craft their work density can be high, but for this to pay off they have to be used intensively. For commercial operations this means just a few ferry operations have really high potential such as Hong Kong to Macao or Guangzhou, the English Channel and Solent, and some routes in

Japan. There are plenty of other routes that can be worked intensely, but the ticket price point is too low. Some services have been able to continue running with SES in the Mediterranean but in most locations that have been tried, the traffic density is not high enough, or the trip length and exposure fits better to a larger catamaran that can carry cars. In the military arena special applications have proven (relatively) economic, such as Mine countermeasures, fast strike craft, and ACV landing craft. Within the envelope that lightweight diesels can power an ACV it has proved practical to design relatively efficient craft for all sorts of utility purposes, and this is likely to continue. As the last generation come up for replacement it should be possible to advance technology to maintain the economy and so attractiveness of the vehicle for these applications. At present China is opening up as an opportunity due to the rapidly developing economy encouraging personal travel for pleasure as well as greater mobility for the population in general.

Performance in a Seaway

The ACV is supported by an air cushion, hovering over the water surface. The air gap between the skirt lower edge and water surface is very small so as to reduce the air flow rate and lift power, a few centimetres average gap. When an ACV is running in waves, the air leakage under the skirt will increase significantly unless the skirt is able to respond to the water surface and maintain the minimal air gap. This was the big challenge for early developers as air jet flexible seals were improved into bag and segment systems.

Following research in the 1970s–1990s designers have developed more flexible skirts to respond to the wave profiles such that "wave-pumping" and "cobble stoning" are less of a problem both for ride comfort and for skirt wear. This is nevertheless still an important area of research and should see further advances in the next decade, along with spray depression systems. This design challenge can be likened to tyre manufacturer research into rubber compounds and tread patterns for dry, wet, off road, snow, ice etc. Plenty of challenge here with ACV running over seas, mud, sand, rock, tundra etc.

In addition, for an ACV running in waves, wave resistance increases significantly due to interaction between the cushion induced waves and the skirt when boating below "Hump speed". What does the "Hump speed" or "Hump drag" mean? Figure 2.66 shows a typical speed vs. drag graphic of an ACV. Point A is the resistance hump just prior to transition between operation in displacement mode, and planing mode.

Once planing the cushion induced wave-making of SES and ACV dies away, so that drag will not increase but drop down, then the craft will accelerate with unchanged engine output, and the drag reduces with increase of speed to point B called the drag hollow, and the craft has "taken-off". After point B drag increases with increase of speed, to point C where maximum thrust power limits further acceleration. It is most important for a well designed ACV to pass through the drag hump

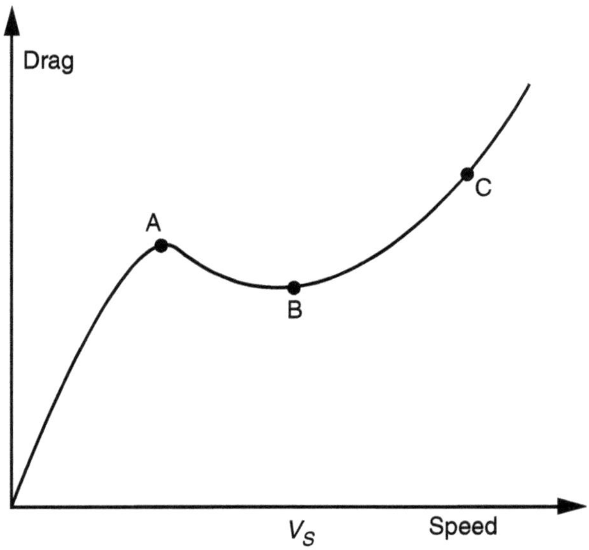

Fig. 2.66 Drag curve of ACV with speed

with a good margin of thrust and transit smoothly to planning operation. Optimisation of the drag curve by attention to the skirt system and plan form geometry is a central part of ACV design, together with design of propulsors that provide high thrust at low speeds for transit through hump, and at the same time high thrust at service speed so as to minimize speed loss in a seaway. It is this design demand that has lead to adoption of ducted propulsors for amphibious hovercraft as these can be designed for high thrust at static and low speeds with smaller dimensions than an open propeller, so making the craft layout more compact. A side effect is a lower noise emission, which is very useful also.

When an ACV hovers statically on the water surface, Fig. 2.67, a depression in the water surface is made under the craft by cushion pressure equivalent to the craft weight. While not filled by a solid hull, the movement of the depression in the water caused by the cushion pressure generates the wave-making drag as speed increases. In addition, the wide frontal area of the wetted skirt in the depression plus the bow wave height generates a resistance, and also a secondary wave-making drag, superimposed on the wave system from the cushion depression. The wave generation builds quite quickly to an initial peak at Fn = 0.35 approx, and another at Fn = 0.63 approx for normal shaped ACV's with L/B between 1.5 and 2.5. The initial peak dies due to wave interference patterns while the principal drag curve builds to the main hump which is also at the speed that planing operation starts, hence the drop away above this until either wind drag takes over, or drag due to waves in the seaway, or a combination.

Designers make every effort to reduce the hump drag by optimisation of skirt configuration, and use of "responsive" flexible skirts; however, it is particularly

Fig. 2.67 Airlift hovercraft P34 hovering static over water

difficult to reduce the drag peak in the case of craft running in a rough sea, so an ACV may accelerate comfortably through hump over calm water, while taking much longer to take-off above hump speed in rough seas. This is a limitation of ACV in open waters. Such hump drag exists on all high-speed craft as they transition from displacement boating mode to planing or flying. The drag peak is not as accentuated for other craft as for ACV's though.

The main opportunity for ACV designers is to minimize the cushion pressure and so the depth of the cushion depression. In the 1970s and 1980s this lead to several craft designs such as the SR.N6 and SR.N4 being lengthened and achieving improved performance with higher carrying capacity.

Size Limitations

Since wave-making drag increases proportional to cushion pressure squared, a large cushion area will reduce wave-making drag. The large dimensions, on the other hand, cause a significant increase in hull weight eating in to this advantage unless very light materials are used, similar to aircraft manufacture. Above craft dimensions of 30 m length, non-metallic materials are prohibitively expensive for all but naval craft and so this really defines the upper limit for an SES unless operational speeds of 40 knots or less are acceptable. This speed range is fulfilled efficiently by modern catamarans and is the reason why SES have not taken off as a large commercial market.

ACV on the other hand are not challenged by competition in their special area of expertise, and so the ACV platform has also been developed to allow movement of very large loads at slow speed over sensitive terrain such as the tundra as well as for the high-speed amphibious marine applications. The overall dimensions of air propellers and so the propulsion power than can be installed provide a limitation to the practical size for an ACV. For example, the diameter of the air propellers on SRN4 (a craft weighing 300 t), driven by four gas turbines with 3,400 hp each, were 6.4 m.

The use of ducted propulsion can alleviate this limit, as is illustrated by the Zubr, with its three large ducted propellers across the stern. This represents the technology limit so far, so a 500 t craft or thereabouts is probably the practical upper limit. Most utility applications for ACV are actually for much smaller craft in the 5–50 t size, and this has been the case for the last 20 years or so now. This contrasts with hovering load platforms that have been designed for payloads up to 2,500 t, but do not have to propel themselves, as they are generally manoeuvred by towboats or tracked all terrain vehicles.

Regarding SES, although high-speed water jets or propellers can be designed to deliver propulsion of many thousands of kilowatts, diesel power is still difficult to design into a large SES successfully, and lift systems, both the mechanical equipment and the cushion seals are complex for both construction and maintenance.

Developing the thicker skirt material necessary for such SES is also a challenge. This is due to high abrasion rate of the skirt material due to high frequency flutter of the segment tips at the water surface at high speed. Engineers working on the 3KSES in the 1970s estimated the life of the skirt lower segments on such a ship would be just two or three round trips in Atlantic Sea between North America and Europe! Materials and design have both moved forward tremendously since then; nevertheless, the segment is a wearing component and has to be designed for easy and minimal cost replacement, just as a car's tyres.

Moving Forward

We shall see in Chap. 6 that catamarans have become the concept of choice for ferry operators seeking high-speed service for passengers and vehicles, and the SES has migrated to the niche of fast military coastal patrol. Amphibious craft are steadily building a market in paramilitary services, and are set to move further into utility services for environments where other craft simply cannot access effectively. The next big challenges for air cushion technology lie in arctic regions, and within the military for amphibious assault. The US Navy has set a challenge for a rather large craft that can be deployed as a catamaran and operated in its assault role as an SES up to shore and transfer to full peripheral skirt to become amphibious for the transition ashore. We introduce and discuss this in Chap. 8. This challenge is certainly a worthwhile one for cushion craft as it should produce a number of spin-offs as well as the craft itself. The arctic challenge is also one focused on cushion and skirt technology, this time the application centred on oil or perhaps energy industry developments in the Arctic Ocean. More resilient flexible materials, and skirt designs reducing air gap and so flagellation damage may be central to success in this market. The improvement of main engine efficiency, and maybe also eventual move to hybrid technologies, should be seen in the coming decade or so.

Some principal dimensions and features of a selection of ACV are listed in Table 2.1 and example SES in Table 2.2 below. Extensive current information is available in the Jane's High-Speed Marine Transportation reference books, see [2-9] and Internet site under resources.

Table 2.1 ACV example data

Name	Country	Year original	Length (m)	Beam (m)	Weight (t)	Cushion pressure (kg/m²)	Power (shp)	Speed max/waves (knots)	Wave height sign't (m)	Passengers/ crew	Cars/cargo payload (t)
(a) Historical military and larger commercial craft											
SR.N5	UK	1966	11.8	7.7	6.7	80	900	60/40	2	20/2	–
SR.N6	UK	1967	14.6	7.7	10.0	95	900	52/35	2	38/2	–
SR.N6 Mk 6	UK	1969	19.2	7.9	17.2	126	1,400	60/50	2	14/2	–
SR.N4 mk3	UK	1970	56.4	23.2	300	257	18,000	65/50	4	450/6	55
N500	France	1974	50.0	23.0	265	315	17,000	70/45	3.5	350/6	45
VT-2	UK	1974	30.2	13.3	105	275	7,600	63/50	3	130/3	3/32
BH-7	UK	1974	23.1	11.3	50	205	3,800	60/50	3	170/3	6/14
Aist	Russia	1976	47.3	17.8	303	360	19,200	70/50	4	200/15	2 tanks
Lebed	Russia	1977	24.6	10.8	114	451	20,000	55/50	3	120/6	/35
JEFF (A)	USA	1975	29.0	14.6	157	400	22,500	60/50	4	/3	6/54
JEFF (B)	USA	1975	26.4	14.3	150	420	22,500	62/50	4	/3	6/54
Voyageur	Canada	1978	20.0	11.2	40.8	190	2,600	50/40	2	/2	/30
LCAC	USA	1980	28.62	14.33	157.4	400	15,820	60/50	4	170 troops	6/54
Griffon 1050	UK	1980	9.5	4.89	2.5	54	140	30	0.45	11/1	–
AP-1.88	UK	1981	21.5	10.1	29	200	1,500	55/40	2	88	–
MV.PP15	Japan	1985	25.1	11.1	50	230	4,400	65	2	70	–
Griffon 2000	UK	1985	12.7	6.1	5.5	70	440	35	0.6	22	–
Zubr	Russia	1990	57.0	21.5	415/555	480	59,180	63/55	4	140/31	3 tanks
Griffon 380	UK	1990	6.8	3.76	1.0	39	116	30	0.3	4/1	–
Griffon 2400	UK	1990	13.4	6.8	6.0	66	585	35	0.7	25	–
Griffon 500	UK	1995	8.04	3.92	1.25	39	84	30	0.4	6/1	–
Griffon 3000	UK	1995	18.4	10.1	10.0	54	590	37	1.0	42	–
ABS M-10	UK	1998	21.6	8.8	20.0	105	1,576	40	1.5	96 (40)	– (2)
Griffon 4000	UK	2000	21.2	10.1	15.5	72	1,180	35	1.0	68	–

(continued)

Table 2.1 (continued)

Name	Country	Year original	Length (m)	Beam (m)	Weight (t)	Cushion pressure (kg/m²)	Power (shp)	Speed max/waves (knots)	Wave height sign't (m)	Passengers/crew	Cars/cargo payload (t)
Griffon 8000	UK	2005	21.3	11.0	25.0	105	1,680	45	1.1	82	–
Griffon 8100	UK	2007	22.6	11.0	30.0	120	2,000	45	1.1	98	–
BHT 130	UK	2008	29.3	15.0	50.0	114	4,800	45	2.0	130	–
BHT 150	UK	2009	30.8	15.0	60.0	130	4,800	45	2.0	150 (47)	– (2)
BHT 160	UK	design	32.3	15.0	64.0	130	4,800	45	2.0	160 (47)	– (2)
BHT 180	UK	design	33.7	15.0	65.0	130	4,800	45	2.0	180	–
(b) Examples of recent small smaller ACV's											
Pacific H/C											
Slider	Australia	2000	4.2	1.9	0.625	80	55	40/30	0.5	1/1	–
Discovery 7	Australia	2003	6.5	2.9	1.9	110	174	37/28	0.5	6/1	–
Explorer	Australia	2005	7.3	3.5	2.7	120	240	35	1.0	12/1	–
Odyssey	Australia	design	10.5	4.5	5.4	120	300	40	1.3	20/1	–
Airlift H/C											
Hoverflyer	Australia	1990	6.34	2.85	1.65	100	135	35/25	0.5	6/1	–
RIVAC	Australia	2006	7.13	3.98	1.7	65	167	40/30	0.5	7/1	–
Wildfire	Australia	2007	7.08	2.7	1.95	110	180	36/27	0.7	8/1	–
Pioneer mk3	Australia	2008	12.15	5.7	4.7	75	330	40/30	0.5	22/1	–
Neoteric H/C											
Neoteric 4	USA	current	4.2	2.5	0.49	52	65	40/30	0.6	3/1	–
Neoteric 6	USA	current	4.7	2.5	0.76	72	100	45/30	0.6	5/1	–
ACV Designs											
Canair 500/2	Canada	current	4.98	2.18	0.60	63	65	30/26	0.4	3/1	–
Universal H/C											
UH 19XR-SAR	USA	current	6.02	2.29	0.94	80	135	48/30	0.2	6/1	–

Table 2.2 SES Example data

Design Name	HM 221 Ferry	HM 527 Ferry	Santa Maria (+12 others) CIRR 120P	Ulstein UT 904	(Hisho, then Kibo) TSL-A Prototype	Gong Yang Gold 37 m SES	Ogasawara TSL-A140 (Techno super liner)
Country	UK	UK	Norway	Norway	Japan	S. Korea	Japan
Company	Hovermarine Ltd	Hovermarine Ltd	Cirrus design Brodrene Aa construction	Ulstein technology (with CIRR + Brodrene Aa)	Mitsui Engineering & Shipbuilding Co Ltd	Samsung Heavy Industries Co Ltd	Mitsui Engineering & Shipbuilding Co Ltd
Delivery year	1982	1983	1988	1990	1994–97	1999	2004–5
Loa, m	21.19	27.2	35.25	39.0	74	36.5	140
Beam, m	5.91	10.2	11.5	12.0	3.5	12.2	29.8
Draught, m							
Hull borne	1.76	2.55	2.1	2.6	3.5	2.2	5.0
Cushion Borne	1.21	1.7	0.7	0.8	1.1	0.8	2.34
Mission	Passenger	Passenger	Passenger	Passenger	Pass + Vehicle	Passenger	Pass + Vehicle
Passengers	112	200	330	320	260	352	740
Vehicle/payload	–	–	–	–	30 cars	–	cargo 210 t
Propulsion	2 GM 8V-92T1	2 MTU 12V396TB83	2 Deutz MWM TBD604B16V	2 × Diesel	2 × GT	2 × MTU 16V396TE74	2 × GT UTC MFT8
Power kW	2 × 363	2 × 1,050	2 × 1,704	2 × 2,000	2 × 12,000	2 × 2,000	2 × 25,180
	2 × FPP	2 × FPP	2 × WJ KMW 56/562	2 × WJ KMW	2 × WJ KMW	2 × WJ KMW 63SII	2 × WJ KMW VLWJ235
Lift	1 Cummins VT-555	1 MTU 6V396TB83	2 MWM TBD234-8V	3 diesel	4 diesel	2 MTU 8V183TE72	4 Niigata 16V20FX
	Centrifugal fans	Centrifugal fans	2 double inlet centrifugal fans	Centrifugal fans	8 centrifugal fans		

(continued)

Table 2.2 (continued)

Design Name	HM 221 Ferry	HM 527 Ferry	Santa Maria (+12 others) CIRR 120P	Ulstein UT 904	(Hisho, then Kibo) TSL-A Prototype	Gong Yang Gold 37 m SES	Ogasawara TSL-A140 (Techno super liner)
Power kW	200	550	2×261	3×380	4×1,500	2×405	4×4000
Skirt system	Loop+finger bow, bag at stern	Loop+finger bow, bag at stern	Full finger (bow), Lobe (stern)	Loop+finger bow, bag at stern	Full finger (bow), Lobe (stern)	Loop+finger bow, Lobe at stern	Full finger bow, Lobe at stern
Structure	GRP single skin	GRP single skin	FRP/PVC Foam Sandwich	FRP/PVC Foam Sandwich	Aluminium	FRP single skin hull FRP sandwich Deck & Super structure	Aluminium
Range, n mile	140	200	340	300	500	250	1,200
Crew	3	3	6	6	6	6	50
Displacement, t	28	87	90	100	2,785	n/a	14,500
Max speed, kt	35	36	50	50	45.23	50	42.8 in 2 m s
Cruise speed	33	32	45	40	40	45	38
Ride control	none	none	Ride control system (RCS) fitted Louvre valves, Pressure Sensor & micropro-cessor Control to reduce vertical Acceleration	RCS fitted Louvre valves, Pressure Sensor & microproces-sor Control to reduce vertical Acceleration	RCS available	RCS available	RCS available
Frl	1.25	1.13	1.38	1.32	0.86	1.36	0.59

Country	USA	China	USA	USA	Russia	Norway	Norway
Company	Bell aerospace	MARIC design Wuhu Shipyard build	Bell Halter	Bell Halter	Zelenodolsk, Kazan	Umoe Mandal AS	Umoe Mandal AS
Delivery year	1972	1984	1983	1986	1988, 1995	1997–2008	2000–2012
Loa, m	23.7	40	33.5	48.7	64	55.2	47.5
Beam, m	10.7	8.2	11.88	11.88	18	13.6	14.3
Draught, m							
Hull borne	n/a	2.45	2.51	2.83	3.0	2.15	2.7
Cushion Borne	n/a	1.85	1.67	1.67	n/a	0.9	1.0
Mission	Prototype test craft	Ferry	Coastguard cutter	Navy demonstar	Guided missile corvette	Mine counter measure	Fast strike craft
Personnel, Crew	4	7	6	6	35	38	14
Mission	6	251 passengers	12	119	68 combat		7
Payload	Trials outfit instruments	Passengers	Coastguard outfit	Demonstration	Military outfit	Military outfit	Military outfit
	10 t				Missiles, Cruise missiles and guns	Mine counter measures	Fast Patrol Craft
Propulsion	3 P&W GT FT-12A-6	2 MWM TBD234-16V	2 DDA 16V-149TIB	2 GM 16V-149TI	2M10-D1	2 MTU 12V396TE84	2 P&W ST18
					GT 2 M511A diesel		2 P&W ST40
Power kW	3×3,340	2×780	2×1,336	2×1,188	2×22,500	2×1,400	2×2,000
					2×7,500		2×4,000
	2×surface ride propeller	2×3b FPP	3b FPP	3b FPP	2 3blade CPP	2×WJ	2 KMW WJ 80 cmØ
Lift	3 UACL GT ST6J-70	1 MWM TBD234-16V	2 DDA 8V92N	4×GM 8V-92TI	2 M52OM3 Diesel	2 MTU diesel 8V396TE54	2 MTU diesel

(continued)

Table 2.2 (continued)

Design Name	HM 221 Ferry	HM 527 Ferry	Santa Maria (+12 others) CIRR 120P	Ulstein UT 904	(Hisho, then Kibo) TSL-A Prototype	Gong Yang Gold 37 m SES	Ogasawara TSL-A140 (Techno super liner)
Power kW	3×370	780	2×260	4×325	2×2,550	2×700	2×735
Skirt system	Bag and Finger at bow, multi-loop at stern	Bag+Finger at bow 2 lobe at stern	Finger at bow Multi-loop at stern	Finger at bow Multi-loop at stern	Finger at bow, Stern multi-loop	Loop+finger bow, stern multi-loop	Loop+finger bow, stern multi-loop
Structure hull, superstructure	Al alloy, Al alloy	Steel, Al alloy	Al alloy welded Al alloy welded	Al alloy welded Al alloy welded	Al alloy welded Al alloy welded	FRP/PVC Foam sandwich	FRP/PVC Foam sandwich
Range, n mile	n/a	200	1,550	2,950	2,500 at 12 knots 800 at 55 knots	1,200	800
Displacement, t	90	123.5	134	207	1,000	367–375	274
Speed, kt Max service	90.3 calm	27.5 24.0	28	28	55 GT power 12 cruise, diesels	20+ transit 5+ hunting, 12+ sweeping	60 45
Service sea state	SS3 GoM	1.5 m seas	SS3 Atlantic	SS4 Atlantic	SS5 Black sea	SS5 Atlantic	SS5 Atlantic
Ride control	none	none	none	none	No details available	Cushion louvre valves, Pressure Sensor & microprocessor control to reduce vertical acceleration	Cushion louvre valves, Pressure Sensor & microprocessor control to reduce vertical acceleration
Frl	3.04	0.71	0.79	0.66	1.13	0.53	1.43

Chapter 3
Wings in Ground Effect: Ekranoplans and WIG Craft

The "Caspian Sea Monster"

During the 1970s, NATO discovered a Boeing 747 size object flying at about 500 kph at zero altitude from their satellite surveillance of Russian military operations in the Caspian Sea. The big surprise was its speed and likeness to large aircraft while skimming the sea surface. NATO called this flying object the "Caspian Sea Monster". Analysis suggested the craft was actually double the size of a Boeing 747, weighing about 500 t, so that it could possibly accommodate about 900 marine troops (Fig. 3.1). It was not confirmed until after the collapse of the Soviet Union at the beginning of 1990s that this craft was the prototype for a military amphibious fast deployment squadron.

Skimming Close to the Surface

Walking along the beach, seagulls can often be seen to skim near the water surface to take food from the sea. They do not flap their wings but are able to glide for quite a distance. The seagulls are using surface effect and get more lift and less resistance while flying in this way.

Aircraft pilots also experience the same phenomenon just before touch-down when landing a low wing aircraft. They feel the airplane pushed up from the ground with extra lift just before the landing gear makes contact and float along until more speed is lost and finally they touch down. It is a "ground effect" that causes the aircraft to glide above the runway until sufficient speed is lost for the cushion to decay. Lift increases and resistance reduces near the ground in the zone less than one wing chord above it—the surface effect zone (SEZ) or ground effect zone (GEZ).

In 1903, the Wright Brothers flew relatively long distances in the SEZ with their biplane, using very little power [3-1]. They were aware of the higher lift forces when gliding close to the ground, but were aiming to fly higher into the air (Fig. 3.2).

L. Yun and A. Bliault, *High Performance Marine Vessels*,
DOI 10.1007/978-1-4614-0869-7_3, © Springer Science+Business Media, LLC 2012

Fig. 3.1 KM

Fig. 3.2 Wright flyer

Later, in the mid 1930s Kaario of Finland started to build and test craft operating in strong ground effect, see Fig 2.17. Kaario's concept was for a high speed boat that could glide over ice as well as water. However, due to greater interest in the aeronautical industry for development of passenger aircraft, floatplanes, and seaplanes, the captured air bubble craft built by Kaario were not developed further. It was 30 years later, at the beginning of the 1960s, that Alexeyev began his development of Ekranoplan's in Russia [3-2]. In the USA, research began a little later in the 1960s targeting the same military objective—the quick reaction amphibious force [3-3]. While the US research went as far as experimental prototypes before being overtaken by the ACV and SES developments, in Russia WIG craft were developed to amazing size by the USSR for full military service in the Caspian Sea.

Basic Principles of WIG Craft

Ground effect is caused by the following physical phenomena.

- Flow blockage between the wing underside and the ground increases the pressure on the lower surface of the wing so as to increase the lift.
- In addition, for a wing operating close to the ground or water surface, the down wash velocity caused by wing tip vortices is reduced. This effect reduces the induced resistance caused by the wing tip induced velocity.

Figure 3.3 shows these phenomena. The upper figure shows the wing tip vortex without ground effect, while the lower figure shows that with ground effect. The ground reduces the airflow down wash velocity leading to a decrease of resistance and an increase in lift, as well as an increase of effective aspect ratio for the wing, ($AR = B/C$, where B is the wingspan, and C the wing chord). Craft designed to fly close to the ground and use this effect are called Wing in Ground Effect craft (WIG), and also Ekranoplan, (the Russian Terminology, since the craft type was first developed in Russia).

The expression H/C represents the WIG relative flying height, where H is the flying height and C the wing chord. Ground effect increases as relative flying height reduces. The zone where $H/C<0.1$ is the strong SEZ. In this zone, the surface effect is relatively strong. When $H/C>1$ the ground effect is very small. To use the phenomena efficiently, a WIG wing should have short wingspan and long chord. This configuration is evident in the KM diagram shown below. Figure 3.4 shows a general arrangement of the craft, in which the main elements are identified, as follows:

1. Eight turbojet forward engines providing ram air lift at low speed and propulsion at high speed
2. Low aspect ratio Main wing
3. Tip plate of the main wing, with buoyancy that stabilizes before takeoff
4. Propulsion engines
5. High positioned horizontal tail stabilizer
6. Elevators
7. Tail Fin
8. Rudder
9. Flaps
10. Main hull/fuselage

During takeoff and touch-down of the KM, jet flow is blown from front mounted jet engines into the lifting cavity formed by main wings and the wing tip plates by

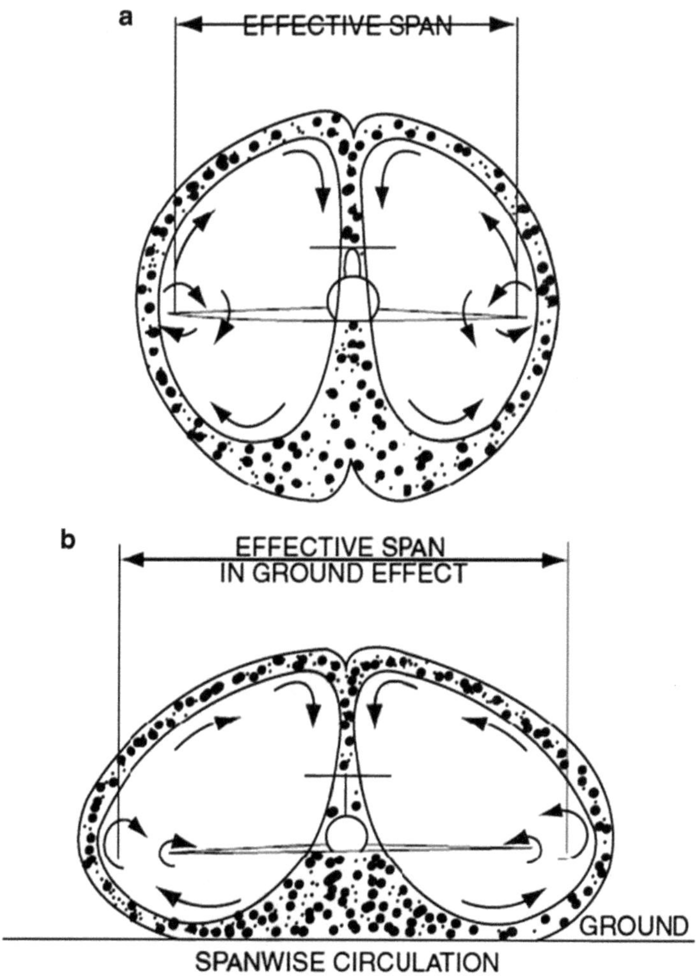

Fig. 3.3 Wing blockage phenomena in ground effect

turning the jet nozzles down, with wing trailing edge flaps also turned down to create just a small gap for air release at the trailing edge of the wings and the ground surface similar to the cushion on an ACV.

Where a WIG is without air jet assistance to create the cushion at slow speed, the planing surfaces of the craft at the underside of the fuselage and side buoys (or wing tip plates) provide planing lift before takeoff, and the wing geometry needs to be arranged to trap air as efficiently as possible. In this case higher water drag is experienced before takeoff and the craft has to be designed more like a floatplane. The craft moving at slow speed also generates a wave making resistance before takeoff, which generates resistance humps similar to that of an ACV for the jet assisted Ekranoplan, or similar to a float plane if the jet assistance is not installed.

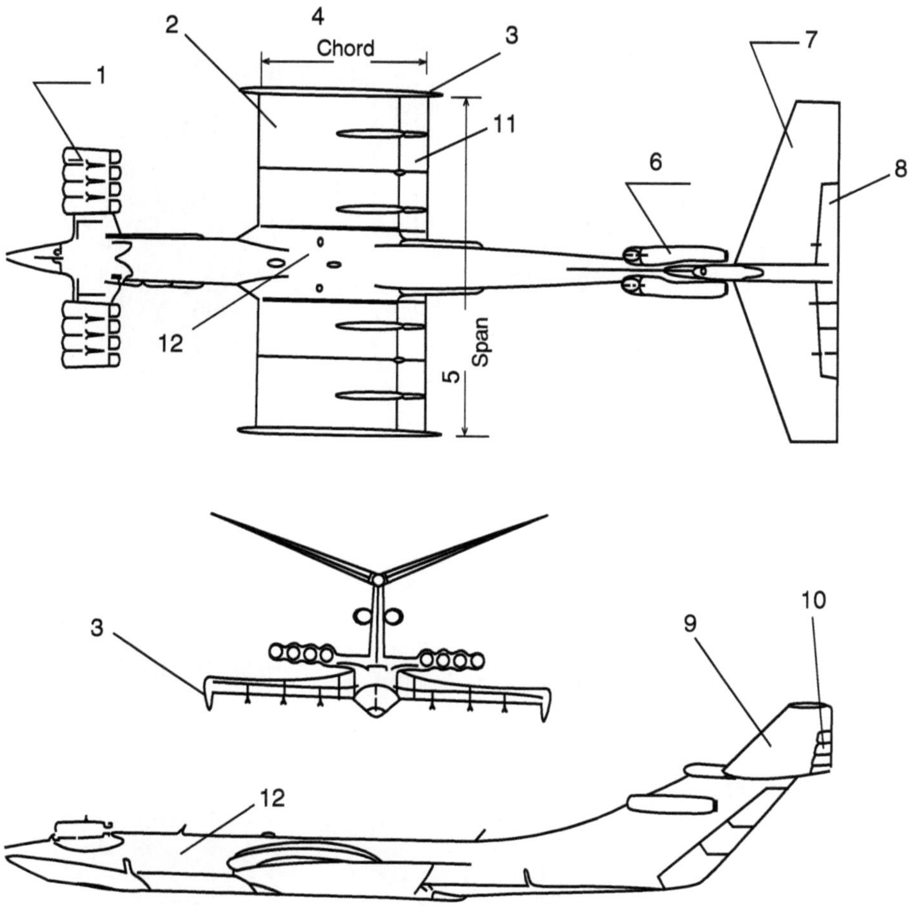

Fig. 3.4 KM 3 view diagram

Figure 3.5a shows the Russian WIG "Orlyonok" during the takeoff operation, and one can see that a wave system and heavy water spray is generated by the cushion pressure under the main wing, similar to an ACV. In contrast to an ACV, after takeoff the wave-making disappears, as in Fig. 3.5b, and only little ripples under the tip plates of the WIG can be seen, generated by the wing tip vortices. At cruising speed the forward lift engines can be shut down. The stern engines propel the craft in cruising flight while tailplane elevators and main wing flaps are used for pitch and roll trimming and dynamic stability. The tail fin and rudder are used for directional control and turning similar to an airplane, though with very shallow banking in turn manoeuvres assisted by the main wing flaps operating differentially. The Orlyonok was the operational design that was developed from the prototype "KM".

WIG have been designed and built in a number of countries as well as Russia, though not as large. There are several different types, so we introduce them and the principles used for each.

Fig. 3.5 (**a**) Orlyonok takeoff. (**b**) Orlyonok in flight

Types of WIG

WIG craft have developed into a number of different configurations dependent on the planned service speed, and the means used to augment lift under the wing at low speed. The original Russian development of Ekranoplan was aimed at fast troop deployment in the Caspian Sea area (and perhaps others if the fuel crisis had not changed the world approach to petroleum in the 1970s). By flying close to the sea surface, the craft could be invisible to radar and too fast for reaction to visual contact.

Development projects carried out in other countries in Europe, notably Germany, aimed at fast coastal patrol at lower cost than helicopters, and a little later at small passenger craft targeting speeds higher than boats rather than as fast as aircraft [3-4]. In Russia, service speed in the 350–500 kph range meant that lifting aids were necessary to augment wing lift at speeds lower than takeoff so as to reduce drag and total installed power, thus the power assisted ram air WIG resulted from this line of development.

The "classic" WIG without powered air jets to augment the natural ram air cushion effect can, however, be successfully designed for service speeds in the range 100–300 kph, and this is where we begin.

Classic WIG

At the beginning of the 1960s, Dr. Alexander Lippisch in Germany developed a prototype WIG, the X-113. The configuration was characterized by the negative dihedral Delta Wing and three axis control surfaces (rudder, elevators and ailerons) similar to an airplane (Fig. 3.6).

The single seat X-113 test craft designed by Lippisch was built under a contract for the German Ministry of Defence by RFB, a company within the VFW/Fokker Aircraft group of Germany and Holland, in 1970. The craft configuration comprises a fuselage with stepped planing lower surfaces, and main wings with significant anhedral and tapered chord so as to create a triangular shaped dynamic air cushion space. Planing floats were mounted on the main wing tips, and outer winglets with 60° dihedral were installed for roll stability in when in flight.

The complete structure of the X-113 was built in composite materials, resulting in a very light weight for its dimensions, of 250 Kg. Propulsion was provided by a single pylon mounted two cylinder Nelson engine of 48 bhp (38 kW) driving a wooden two blade open propeller. This type of WIG uses its wing configuration to

Fig. 3.6 Lippisch X-113

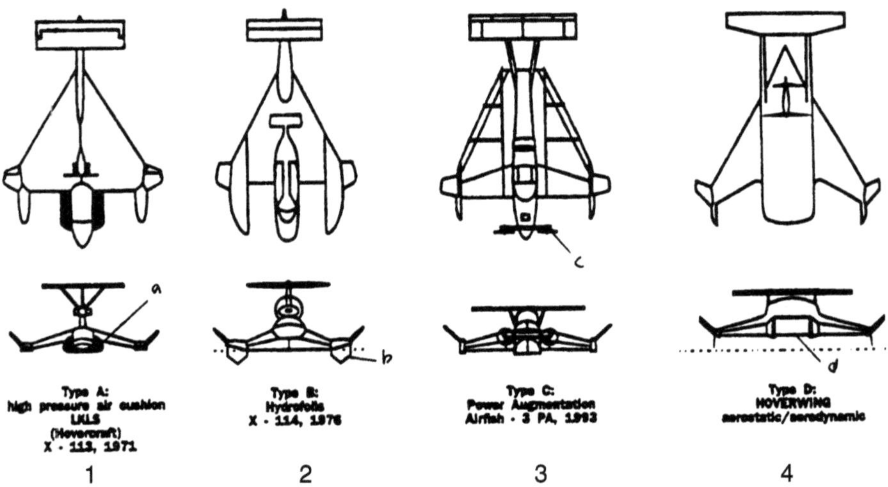

Fig. 3.7 Lippisch X Series WIG development diagram

trap air under it as it moves forward. The cushion effect supports only part of the craft weight until takeoff speed is reached, so drag from the fuselage planing surfaces and two wing floats is quite high at low speeds, and takeoff speed is also high. The negative dihedral tapered wings give static stability both before takeoff and afterwards as the centre of area and the centre of lift are close. In contrast the rectangular plan form of the KM Ekranoplan wings meant that the bow thruster's angle and also the wing flap angle had to be carefully controlled during takeoff to cruise, and back down through cruise to ram cushion operations. The Lippisch form avoids most of this, with the benefit that the structure and control systems are simplified compared to craft such as the Russian Ekranoplan.

Figure 3.7 shows the development of the X series, with different takeoff aids. The prototype of this series, X-113, was tested with a cushion like system under high pressure around the bottom of fuselage, to form a ground effect air cushion. It was able to fly in ground effect, and also as a light aircraft up to 800 m altitude (accompanied by a Bell Huey helicopter escort!) in trials during 1971 and 1972. The X-113 took off at 40 kph and cruised at 80 kph.

Figure 3.7 (2) next shows derivative X-114, with takeoff aided by hydrofoils, The leading particulars of X-114 are length 12.8 m, wingspan 7 m, height 2.9 m, takeoff weight 1,500 kg, engine 147 kW mounted at stern, max speed 200 kph, with 6 passengers. The cruise speed was 150 kph and its takeoff speed was 100 kph. Initially the craft was fitted with a wheeled undercarriage to allow it to drive up a concrete launch ramp from the water. The hydrofoils on X-114 shown in Fig. 3.7 were an experiment intended to shorten takeoff but were found to cause problems particularly on landing since unless a positive angle of attack could be maintained as the foils re-entered the water, they would create downward pulling force dragging the side buoys into the water very quickly. Prior to takeoff, the foils lifted the craft effectively but reduced the ground effect and so actually increased the speed necessary for takeoff—the

Fig. 3.8 X-114

Fig. 3.9 Airfish 3

opposite of the intention. In Fig. 3.8 we see the X-114 in flight without the hydrofoils. X-114 completed its trial programme with the German Ministry of Defence very successfully, but did not result in a deployment plan for operational craft.

Two of the key personnel in RFB who were involved with the X series of craft, Hanno Fischer and Klaus Matjasic, formed their own company in the late 1980s with the intent to progress the WIG craft designs. They produced three successive experimental craft, the Airfish 1, 2, and 3. Figure 3.9 shows the Airfish 3 in flight. The craft were first produced with a development of the wing geometry from RFB as can be seen in Fig 3.7, Type C. Propulsion of all these was similar to the X series. They had ideas to use the forward mounted propeller lift augmentation as shown for type C but moved from this to a different idea which they called the hoverwing. The Airfish and Hoverwing series were all craft having flying speed in the range of 120 kph.

The hoverwing principle is shown in Fig. 3.10, where an air scoop in the propulsion propeller pylon is used to feed air to a cushion below a catamaran shaped hull. Figure 3.11 shows the Hoverwing 80 flying. At slow speed the cushion air inflated flexible seals to provide support for the craft and minimize drag. Once flying speed

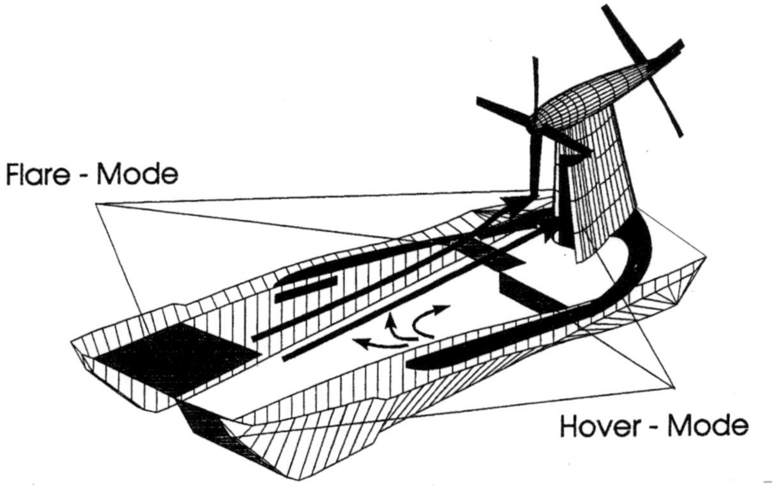

Fig. 3.10 Hoverwing diagram

Fig. 3.11 Hoverwing flying

was reached the seals would flatten against the hull and allow the airfoil shaped cabin to provide lift as well as the outer wings. This geometry, type D in Fig. 3.7, also had the characteristic Lippisch plan form which keeps the centre of lift close to the same position whether in flying mode or in ground effect cushion mode, thus ensuring simple flying control.

An alternative to the single wing aircraft plan form was developed by Dr. G.W. Jörg in Germany based on his experience as a pilot of nearly 20 years during the

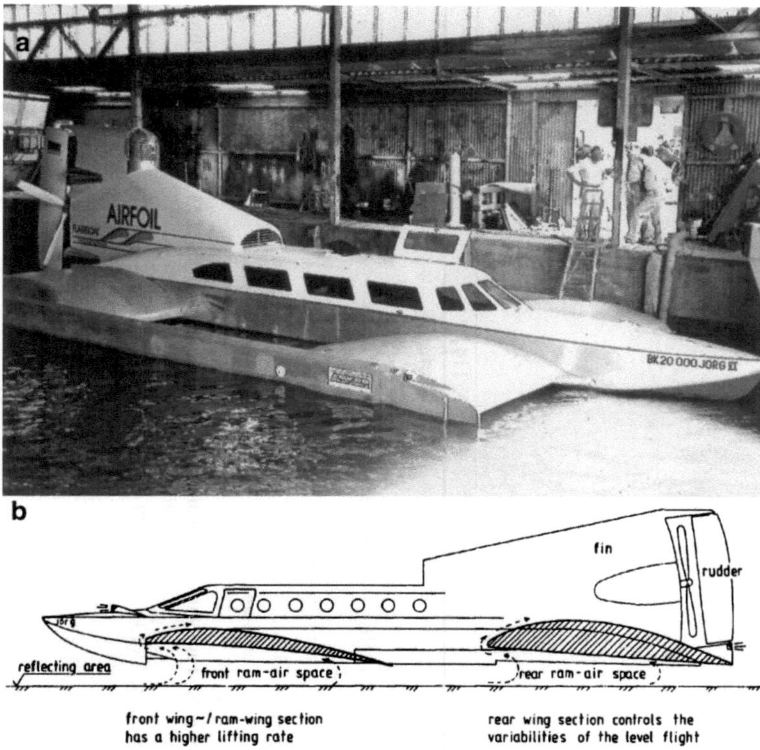

Fig. 3.12 Jorg Craft (**a**) Jorg TAF 8-4 prototype afloat in boathouse, (**b**) tandem wing cross section of Jorg TAF 8 WIG craft

1970s and 1980s. His aim was self-stabilized flight in the strong SEZ so as to build craft to marine regulations, since at that time there were no rules available for WIG other than to follow either aircraft regulations, the approach followed by Lippisch and RFB. Dr. Jörg began with model experiments using similar geometry to the Lippisch design and moved on to using two large identical parallel wings in a tandem arrangement, positioned in line to provide self stabilized flight. There were no pitch and roll controls, since it skimmed in strong SEZ. His craft used a ram air cushion underneath the wing as shown in Fig. 3.12.

The prototype craft were successful in achieving their design target whereby flight is stable with the side buoys skimming the water surface with a clearance of 0.3–0.5 m. They were approved by the German Ministry of Transport under their marine regulations and also under the IMO regulations for motor ships in 1994. The TAF craft showed that speeds up to 170 kph could be achieved safely and simply in calm conditions without the complicated power assisted takeoff of the Russian Ekranoplan. Unfortunately, the craft were not attractive enough for commercial or paramilitary missions to be achieved. The main limitation is the sea state suitable for operation in the strong SEZ. The leading particulars of Jörg TAF VIII-4 are given in Table 3.1 at the end of this chapter. His other craft can be found in [3-4].

Table 3.1 Key data for example WIG

	Orlyonok	KM	Lun	Volga-2	X-114	TAF-VIII-4
Country	Russia	Russia	Russia	Russia	Germany	Germany
Type of craft	PARWIG	PARWIG	PARWIG	DACC	Classic	DACC
Mission	Transport	Experimental	Guided missile	Passenger	Passenger	Passenger
Max TO weight, t	140	544	400	2.7	1.5	4.6
Payload, t	20	130	100	0.7	0.38	12 passenger
Main dimension, L × B × H, m	58 × 31.5 × 16	92.3 × 37.6 × 22	73.8 × 44 × 19	11.6 × 7.6 × 3.7	12.8 × 8.7 × 2.9	18 × 7.19 × 4.05
Power plant, PAR thrust, each, t	2 NK-8-4K Turbojet, 10 t	8 VD-7-NM Turbojet, 11 t	6 NK-87 Turbofan, 13 t			
Cruising power, type	1 NK-2MK Turbo-prop	2 VD-7KM Turbojet	2 NK-87 Turbofan	Light aircraft engines	Lycoming IO-360	Marine power 7.8 litre V8 engine
Power, shp	15,000	Jet thrust2 × 11 t	Jet thrust 2 × 13 t	2 × 120	250	750
Cruising speed, kph	375	500	450	120	150	160–170
Range, km	1,300	2,000	4,000	300	200	200
Wave height takeoff, m	1.5–2.0	2.5–3.5	2.5–3.5	0.5	0.5	0.5
Wave height, cruising	No limit	No limit	No limit	0.3	No limit	0.5
Jump	Yes	No	Yes	No	No	No
Jump altitude, m	Up to 50		Up to 50		>100 m	

	XTW-IV	AF3-A	SWAN DACWIG	Ivolga, EK-12 DACWIG	HP-7	FS 8
Country	China	Germany	China	Russia	USA	Australia
Type of craft	Classic	Classic	DACWIG	DACWIG	Classic	Classic
Mission	Prototype	Prototype	Prototype	Prototype	Ferry	Ferry
Status	Under test	Under test	Under test	Under test	design	certified
Max T.O weight, t	6.0	0.7	8	3.7	1.67	4.75
Payload	20 pax	2 persons	20 pax	12 pax	7 pax	8 pax
		0.128 t		+2 crew	0.7 t	0.65 t
Length, m	21.7	9.9	19.04	15	10.67	17.45
Span, m	14.5	8.6	13.4	12.5	7.61	15.6
Height, m	5.0	2.6	5.2	4.7	n/a	4.1
Power plant engines	2xPT6A-15AG	BMW 1200 a/c	2xHS6A + 1xHS6E	2x3M3-4062.10 petrol	n/a	1xGM v8 w/c petrol
Power, shp	1,000	90	2×350	2×150	300	450
			1×300			
Cruising speed, kph	150	120	139	180	200	158
Range, km	500	300	300	2,000	700	550
Wave height, TO and cruising	SS2	SS2	SS2	0.5	0.5	0.5
	SS3	SS3	SS3	SS3	SS3	SS3
Flying height, m	0–2.0	0–1.0	0.8	0–3.0	3.0–5.0	3.0
Jump altitude	n/a	n/a		>40 m	n/a	up to 10 m

Power Augmented Ram Wing in Ground Effect Craft

The Russian "Caspian sea monster" and "Orlyonok" are power augmented ram wing in ground effect craft (PARWIG). Jet flow is blown from the forward engines into the cavity formed by main wing, tip plates, and ground, thus augmenting the lift force at slow speed to support the craft weight. The cushion lift enables the takeoff and touch-down speed to be reduced for a given wing area on the PARWIG so that a craft can be designed for much higher cruise speeds. Rotation of the jet efflux of the forward "lift" engines provides the power to accelerate to the higher cruise speed if required.

From basic theory, a dynamic cushion is easy to form with a low aspect ratio rectangular main wing and side plates. In addition, a rectangular tail wing to control trim and pitching is positioned high so as to avoid the vortex from the main wing and be large enough to ensure longitudinal stability in free air flow. In general this means a tailplane almost as large as the main wing for a WIG so as to be able to develop pitching moments from free air above the SEZ that will control the craft attitude. This generated the craft configuration of both KM and Orlyonok.

The PARWIG concept was the idea of Alexeyev, chief designer of the bureau that first designed hydrofoils for ferry transport on the inland waterways of the USSR (see Chap. 5). The concept came from a mirror image analogy of the shallow submerged hydrofoils. Travelling at 100 kph, a noisy craft might not be stealthy enough, but cruise speeds in the 250–400 kph range—aircraft speeds—certainly would give the advantage of surprise.

The Russian name for the "Caspian sea monster" is "KM" (an acronym standing for experimental model), Fig 3.1. Its leading particulars are as follows: length 92 m, span 37 m, height 22 m, weight 544 t, cruising speed 450 kph, average flying height 4 m, accommodation capacity 900 marine troops. Following the "KM", Russia had a whole series of prototype craft testing design options before selecting the configuration for the smaller troop transport craft "Orlyonok", Fig. 3.5, with length 58 m, span 31.5 m, height 16 m, weight 1,120 t, bow engines: two bow mounted NK-8 turbojets with thrust 10,500 kg each for both lift and propulsion, and a stern propulsion engine: a single NK-12 turbine driving a pair of contra-rotating propellers.

Orlyonok (Sea Eagle) had a cruise speed of 350 kph and a range of 1,000 km at 2 m flight height, accommodating 140 marine troops and the vehicles that they needed. There were three operational craft built and put into service in the 1980s at a base in the NW Caspian Sea. Operations were curtailed when the USSR broke up and military priorities changed for Russia.

Figure 3.13a shows another PARWIG, the "Lun". This was a guided missile WIG weighing 400t with three pairs of launchers arranged on the upper deck, cruising speed 450–550 kph, with eight NK-87 gas turbine engines for both lift and propulsion mounted at the bow. Both Orlyonok and Lun had a large planing plate under the central fuselage that was adjustable in angle by hydraulic power. This helped takeoff and touch-down to be clean, since they functioned as a planing step (see Chap. 4), and the plate was retracted after takeoff to minimise drag.

Having gained operational experience with Orlyonok at smaller scale, the design of the Lun could adapt this to the basis taken from KM to deliver a really powerful attack craft for heavier weather in excess of sea state 3 as well as have an amazing range of 2,000 km—truly a potential master of the Caspian Sea. Figure 3.13b shows a general arrangement of the craft, and Fig. 3.13c shows the craft under construction.

The problem with these very large craft was that they were expensive to operate and so were dedicated to actual military exercises; therefore, the smaller SM-6 (small scale KM built 1967) and SM-8 (small scale Orlyonok built 1972) could be used for inexpensive training purposes. Meanwhile, the attention of Alexeyev's team turned to smaller craft suitable for commercial purposes. Experimental craft SM-9 and SM-10 were built using the same layout and power as eventually used for the Volga-2 (1986) and later the Strizh craft (1991). The Strizh was used as a training craft, as it was designed to reach the weak SEZ, while the Volga-2 was designed purely as a strong SEZ skimmer, similar to the aims of the TAF craft in Germany,

Fig. 3.13 (a) Lun flying and firing missile

b

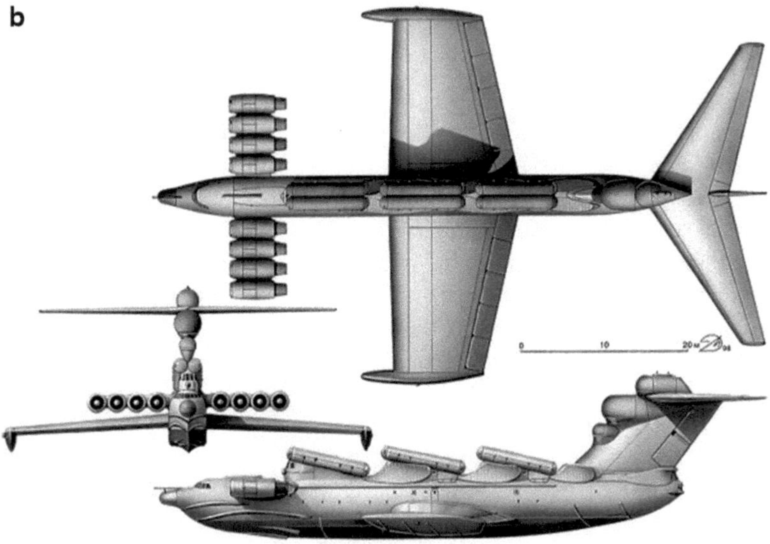

c

Fig. 3.13 (continued) (**b**) Lun 3 view. (**c**) Lun Construction

while using forward mounted ducted twin propellers with guide vanes that could be rotated to give PARWIG effect for takeoff at 70 kph or so and cruise at 140 kph.

Figure 3.14 shows the "Strizh" flying at high clearance (a), and closer to the surface (b). The craft AUW is 1.6 t, max speed 175 kph, accommodating 2 persons, 2 engines with 160 shp each for both lift and propulsion. The Strizh has forward mounted propellers that can be orientated to blow air down under the wing for takeoff or more horizontally for flight.

Fig. 3.14 (**a**) Strizh flying low. (**b**) Strizh climbing

In China PARWIG have also been developed and built as test prototypes. Figure 3.15a, b, shows the Chinese "XTW" series of PARWIG, developed by China Ship Scientific Research Centre (CSSRC), in which (a) is XTW-1, a prototype of this series, and (b) is XTW-3. The aerodynamic configuration is similar to Lippisch's X series with reverse triangle main wing accompanied with high flat tailplane; however, the takeoff capability has been improved due to using a PAR arrangement.

Two aviation piston engines of 220 kW each are arranged in front of the main wing of XTW-3 and give PAR effect during takeoff. The cruise speed is 144 kph and range 500 km. The craft weighs 4,000 kg, with length 17.9 m, with span 12.1 m, and accommodates 12–15 persons. It can fly in SEZ at 0.5–2.0 m clearance, and also as high as beyond the SEZ, with fine stability, just as the Lippisch X series does. What these craft do not have is amphibious quality, so they must start from afloat, and be towed on shore using undercarriage wheels extended.

Fig. 3.15 (**a**) CSSRC XTW-1. (**b**) CSSRC XTW-3

Dynamic Air Cushion Craft

During the 1980s, the chief designer of "KM", Mr. Alexeyev, after development of the military PARWIG craft "Orlyonok", "Lun", etc. realized that the high expertise required to pilot such high speed craft safely, particularly when flying in weak SEZ, i.e. $.(0.1-0.3) < h / B < (0.3-0.8).$, where h flying height, B span of main wing, between the strong SEZ and out of SEZ, was not practical for a commercial craft. He worked with the bureau staff to develop a simpler and more cost-effective commercial WIG, named "Volga-2".

A WIG has high stability when flying in strong SEZ due to the surface effect, and also good stability when flying out of SEZ as an aircraft as long as it is designed with an effective tailplane and suitable centre of gravity for free flight.

Fig. 3.16 Volga 2

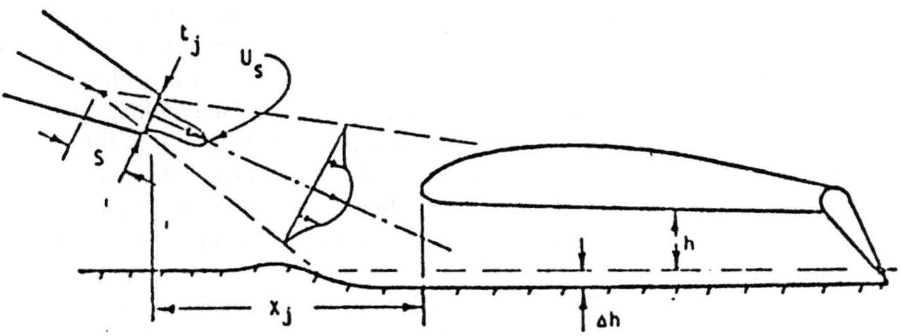

Fig. 3.17 DACC/PAR diagram

However, WIG can have weak stability when flying in the upper SEZ or so-called "transition zone" due to the change of aerodynamic "Pitching centre" and "Heaving centre" with change of flying altitude and pitching angle which is characteristic of WIG. The "KM" capsized while on test in 1970s due to a wind gust lifting it suddenly into the weak SEZ followed by pitch down due to the controls not being able to counter the pitching quickly enough [3-4].

Figure 3.16 shows the "Volga 2" skimming over the water surface. The craft has a medium service speed at 120 kph, low flying height at 0.2 m, and with aid of inflated flexible bag skirts under the side buoys it is amphibious, being able to run on shallow water, ice/snow surface, swamp, and other surfaces where conventional vehicles are unable to excess, just like an ACV. This is due to "dynamic air cushion effect" creating the lift under the dynamic air cushion craft (DACC's) large area wing at relatively slow airspeed.

The aerodynamic rationale of PAR and DACC is similar, as Fig. 3.17 shows. In each case, the air jet from the forward mounted engines blows into the cavity formed by the main wing and flap and the supported surface, to make an air (or jet) cushion to support the craft. The difference between the PAR and DACC is that the former only provide partial lift for improving the takeoff and touch-down capability and cannot support the whole weight of the craft on a cushion at low speed. The main feature of PARWIG is very high speed, as the mission of such craft required, so the aerodynamic configuration does not provide sufficient wing area to create full cushion lift for landing and amphibious quality at low speeds.

The aerodynamic configuration of the slower DACC does provide sufficient for creating an air cushion at low speeds, enabling it to be amphibious and support landing without installing a landing carriage for over ground manoeuvring. There is enough pressured air feeding the cushion area due to the long wing chord, a closable flap for sealing the cushion at the stern, and propeller guide vanes than can be turned down for feeding the pressured air directly into the cushion.

However, such craft also are different from an ACV. An ACV is supported on its air cushion by a "static" air cushion formed by a fan pumping air under the hull and contained by flexible skirts all round, but DACC are hovered with a combination of static air cushion and dynamic air cushion lift. When flying fast, after takeoff, the pilot gradually turns open the wing flaps and turns up the guide vanes so as to increase forward thrust for cruise on full dynamic lift rather than the thruster forced air cushion.

The advantages of such craft are amphibiousness and land-ability, operating clear from the water surface, but still using a small flying height. It is simple and cost-effective, similar to other fast marine craft. On the other hand, the concept does have disadvantages, including limited seaworthiness due to the small flying height, low aerodynamic efficiency due to the low main wing aspect ratio, and lower speed compared with PARWIG.

The Volga-2 has been further developed into a commercial design called the Aquaglide-5, built by the Arctic Trade and Transport Company (see resources). This five seat craft has been sold in significant numbers both in Russia and in places such as the Caribbean as a pleasure craft and water taxi. The Aquaglide has a forward mounted Mercedes 240 kW V8 engine that drives the two rotatable free air propellers via shafts and gearboxes. It has a range of 350–450 km and a cruising speed of 150–170 kph (80–90 knots). The single engine in Aquaglide is more efficient than the two units used for the Volga-2, so operation is rather more economical (Fig. 3.18).

In recent years, the "Ivolga" has also been developed in Russia using a similar power of 300 shp from two engines driving forward mounted ducted propellers that have rotatable guide vanes. The Ivolga is designed as a PARWIG with increased cushion effect at zero speed so that it can "hover" at slow speed over ice or snow—a DACWIG—which we describe more fully in the next section.

So far, now we have followed the concept development through designs for very high speed—almost jet aircraft speeds in fact with the Ekranoplan, reducing down as far as 100 kph cruise speed for some commercial targeted craft. Using forward mounted propulsors and the Lippisch wing form or similar to it, a craft with inherent dynamic stability in skimming or medium SEZ clearance flight can be designed. The next step is a stronger cushion system for amphibious quality to do away with any need for undercarriage.

Fig. 3.18 Aquaglide

DACWIG, the Amphibious WIG

The DACWIG is a derivative both from DACC and PARWIG [3-5]. The purpose is to create a WIG possessing both amphibiousness and higher flying altitude than a DACC. Figure 3.19 shows a three view of DACWIG, type "Ivolga", developed by Russian WIG company, "TREK" LTD. In this figure are shown the following:

- Rotatable forward mounted ducted air propellers, driven by an automobile engine, which can be rotated downwards for low-speed operation and to improve takeoff both for calm water and in waves, and rotated up for flying mode
- Main wing with large plan area and low aspect ratio for an efficient static air cushion
- Composite wing (outer wings) with high aspect ratio for aerodynamic efficiency in flying mode
- Ailerons for strong transverse stability and turning ability
- Planing bottom surfaces with transverse step on side buoy bottoms, for good takeoff capability

Figure 3.20a shows the craft flying at high speed; Fig. 3.20b moving on ice and snow; Fig. 3.21 shows retracted composite wing for water or over ground slow speed manoeuvring. The leading particulars of this craft are listed in the Table 3.1 at the end of this chapter.

The Fig. 3.22 shows another DACWIG—the "SWAN" that was designed and built by the Marine Design & Research Institute of China (MARIC) in Shanghai

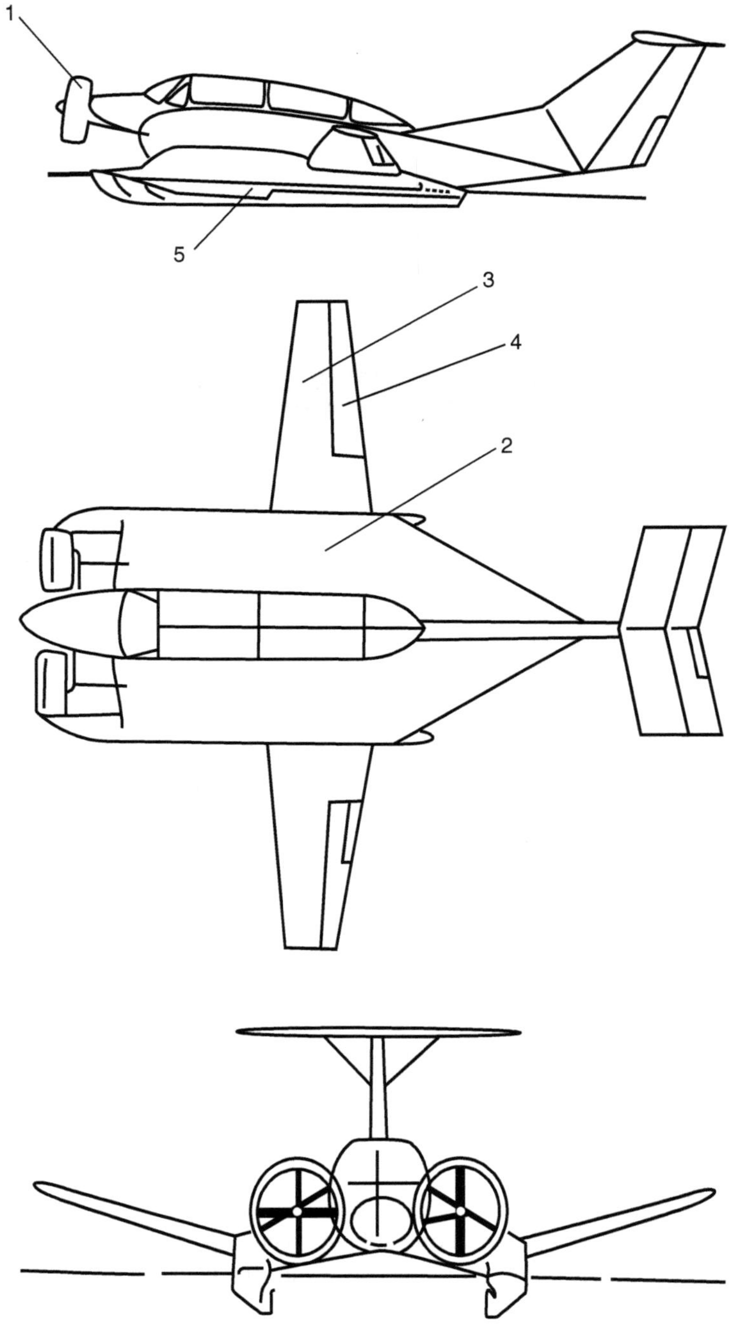

Fig. 3.19 Ivolga 3 view

Fig. 3.20 (**a**) Ivolga flying. (**b**) Ivolga over ice and snow

Fig. 3.21 Ivolga with wings retracted

Fig. 3.22 (**a**) Swan flying. (**b**) Swan landing

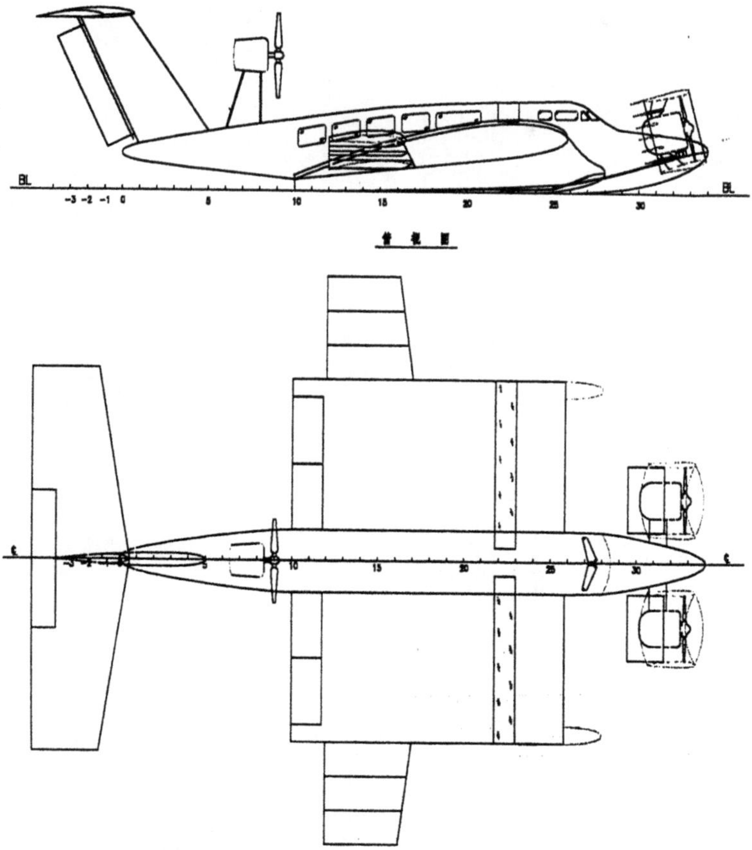

Fig. 3.23 General arrangement of DACWIG swan

(a) flying, and (b) landing [3-6]. It also has bow mounted propulsion engines with propellers mounted in fixed ducts, and guide vanes to direct airflow aft for flying, or below the wing for speeds up to takeoff. Figure 3.23 shows a general arrangement of the "SWAN"

The Swan has larger side buoys than craft such as the XTW series, aimed at full hovering capability at zero speed, and an efficient cruising speed in the region of 120 knots.

Development of a craft such as the Swan demands a combination of aerodynamic and hydrodynamic skills so as to find the right balance between performance at low speeds when leaving or returning to a base location, and its flight performance over a seaway. MARIC carried out a series of wind tunnel tests, towing tank tests, and free flying model tests, to determine both the aerodynamic parameters of

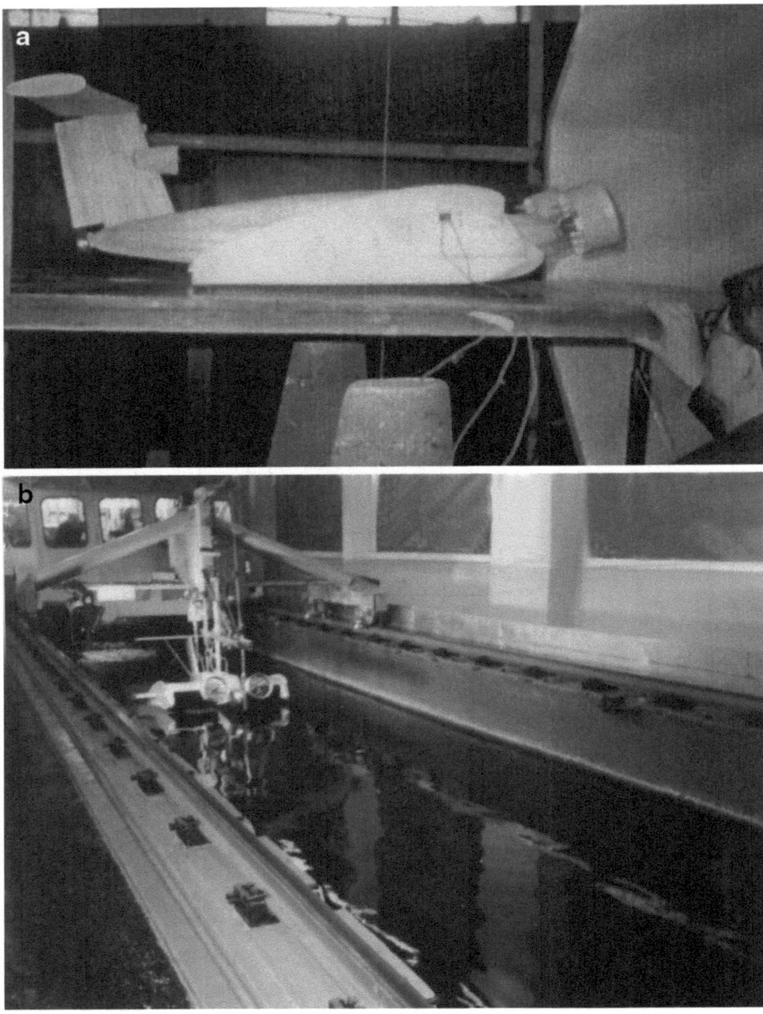

Fig. 3.24 (**a**) Aerodynamic tests. (**b**) Hydrodynamic tests

lift and drag when flying, hovering, and boating. The performance of the ducted bow thrusters was determined through use of a smaller scale prototype craft designated the "750".

Figure 3.24 shows the model tests of "SWAN"; (a) wind tunnel test for aerodynamic experimental investigations; (b) towing tank testing.

Figure 3.25 below shows prototype craft "750" under test. Here, one can see clearly the bow air ducted propeller blows directly the pressured air into the air cushion under the main wing.

Fig. 3.25 The Prototype Test Craft 750

WIG Stability—Personal Experiences

The longitudinal stability of a WIG is low when flying in the transitional SEZ, i.e. when flight height h/c is between 0.5 and 1.0 (where h is altitude, and c wing chord), so pilots have to be careful when handling craft in this region, particularly when flying lightly loaded into a head wind.

The DACWIG "SWAN" is a dynamic air cushion WIG with normal flying altitude about 0.5–0.8 m or h/c in the region 0.1–0.3. Figure 3.26a shows the SWAN in 1997, in its original form, with horizontal outer winglets as it leaves its landing pad at Din Sah lake on air cushion. Figure 3.26b shows the craft in its Mark 2 form with outer wings having dihedral and flow fences.

In the summer of 2000, the craft ran on the Din Sah Lake with speed of about 138 kph at light loaded condition, and experienced a sudden head wind gust. The craft rose out of the strong SEZ and developed a large bow up trim. The running attitude at this time was just like a rattlesnake with head up to air. It was a significantly dangerous motion, and made the craft trim upward further towards stalling. Fortunately, the pilot pulled back the throttle and moved guide vanes behind the propeller downwards in a timely manner to create stronger air cushion and move CP towards the stern, so the craft flattened and landed on the water surface rather rapidly. Nobody was injured or equipment damaged. There were some important visitors on board at the time, however, and they were so scared by the event that this demonstration flight really did affect the development of the WIG. Therefore at MARIC we established a principal rule that 7°deg should be the maximum trim angle for WIG during the operation, particularly in the case of light loaded and head wind condition. Figure 3.27 shows a free flight model of Swan under test at a bow up trim.

Fig. 3.26 (a) Swan at the side of Din Sah Lake. (b) Swan in mark 2 form

Classification of WIG

A key issue for WIG craft has been internationally recognized standards for their construction and operation. Military craft such as those in Russia and prototypes such as the craft built in Germany and China do not need such regulations, though the existence of standards can help a lot. Agree regulations are essential for commercial craft operations. Until the 1990s, WIG craft had to be certificated under aerospace regulations, or in the case of DACC under marine regulations. The IMO were requested to prepare international standards at this time, and so through an international committee produced the interim guidelines which have subsequently

Fig. 3.27 Free flight model of DACWIG "Swan"

updated and advice for operations also added [3-7]. The guidelines for WIG craft design link to the IMO code for high speed craft [3-8].

The Interim International Code of Safety for WIG Craft, issued by the International Marine Organization (IMO) 2002, classified WIG into three types as follows:

"A" Craft not capable of operation without the ground effect;

"B" Craft capable to increase its altitude limited in time and magnitude outside influence of the ground effect in order to over-fly a ship, an obstacle or for other purpose. The maximal height of such an "over flight" should be less than the minimal safe altitude of an aircraft prescribed by International Civil Aviation Organization, ICAO;

"C" Craft capable to takeoff from the ground and cruise at an altitude which exceeds the minimal safety altitude of an aircraft prescribed by ICAO.

Thus, DACC, and DACWIG belong to type "A" (some also belong to B, such as Ivolga), classic ordinary WIG and PARWIG belong to type "B", and seaplanes and flying boats belong to type "C". The International Maritime Organization (IMO) and International Civil Aviation Organization (ICAO) are responsible for jointly setting up the safety code of such craft operating in various operation modes on an international basis, with IMO responsible for type A and B, and ICAO rules to cover type C while in "free" flight. Since IMO will be responsible for all WIG mentioned above, WIG can be designed as a marine vessel, rather than an aircraft, so as to build with lower equipment and systems redundancy requirements and so also lower cost.

The regulations now in force enable any member country to certificate WIG craft or services. The actual implementation of the regulations nevertheless requires a new expertise to be available in the maritime regulation department of a country. It has taken some years for this to be achieved, and in May 2010 certification of an

Fig. 3.28 Wigetworks flightship FS-8

FS-8 12 passenger WIG for ferry operations was completed with the craft officially registered on the Singapore register of ships for operation by a company Wigetworks Pte Ltd in Singapore (Fig. 3.28). The craft operated by Wigetworks was built to Lloyds Register classification notation 100 A1 WIG MCH, Type A, Passenger, WTL 0.5/WEL2.0, WSC 25. A service in Alaska has also gone through certification in 2009 so as to allow an operator to start operations there. Earlier, applications had been made for a service in the Maldives, and approval had been reached in Australia for the Flightship FS-8 to operate and start pilot training. This is now available in Singapore. A body of experience is thus slowly building so as to allow WIG technology to be used for commercial purposes.

The Technical Challenges of WIG Development

What then are the main challenges to be overcome for WIG craft to become commercially attractive, whether for military or commercial missions? The following gives some thoughts.

Takeoff in Waves

Although WIG sea-keeping, particularly for class "B" craft is fine for flying above the sea surface giving almost "unlimited" seaworthiness, WIG takeoff in waves is still a design challenge essentially requiring a sheltered area so that wave height is limited.

Using PAR or DAC technology improves takeoff in waves; however, waves may impact forward mounted air propellers in extreme conditions, as has happened for some real WIG operating in waves during the takeoff. This issue is a matter of craft size and overall configuration. If one considers the configuration of the 750 and

Fig. 3.29 Boeing concept ultra large WIG "Pelican"

scale this up, provision of air cushions in a catamaran configuration may well be advantageous for large commercial craft. Swan has proven the concept of zero speed air cushion works, as has the Ivolga at smaller scale.

Economy

High aerodynamic efficiency may be available on the well-designed WIG operating in strong SEZ, the lift/drag ratio reaching as high as 15–20 when close to the ground or calm water; however, in practical terms, in order to maintain craft seaworthiness, a WIG has to be able to maintain a certain minimum flying height above waves, so decreasing aerodynamic efficiency significantly.

Operating the WIG in medium SEZ might be practical for large WIG without losing too much efficiency due to large wing chord dimensions relative to wave height, and this was the initial impetus behind the USSR development of rather large Ekranoplan at a very early stage. It is also the idea behind the "Pelican" proposed by Boeing as a military logistics transport concept. Figure 3.29 shows an artist impression of "Pelican", which has a wingspan of 152 m, fuselage l×b×h,

122×15×8 m with payload capacity of 1,300 t. Pelican is basically an aircraft designed for low altitude flight, with a range of up to 18,000 km in surface effect at 240 knots or 12,000 km at up to 6,000 m altitude over land. The main wing root chord of approximately 40 m means that while cruising in ground effect it would have an H/C of 0.15–0.4 which would enable efficient flight, and a clearance sufficient for transoceanic transit in the equatorial strip.

The aircraft is designed for landing at major airports and so has a 76 wheel landing carriage. The payload is huge, comparing to the Ekranoplans KM (531 t) and Lun (984 t) and even the catamaran JHSV discussed later in this book (1,500 t). However, Pelican would need large funds for development, and a very strong financial base for dealing with many practical problems, and strong confidence for the developers! At present, the initially proposed mission—delivery of logistical equipment between continents to established airport facilities at much higher speed than present is not a base mission challenge for the US Armed forces. The key challenge is support to large scale amphibious forces and access to unprepared coastline or limited port facilities.

The WIG cannot currently compete with available commercial aircraft types delivered by Boeing and Airbus operating to established airports. The possible way to develop the practical and commercial WIG is to work on the PARWIG as a passenger craft operating at shorter range along the coastal routes between islands and coastal cities, where airports are not convenient for commuting—a modern version of seaplane transport at much larger scale for freight or passengers, using ground effect to improve economy.

Finance and political influence is also needed to set up sufficient infrastructure for landing and terminal operations at each end of a route, together with organizations and air control systems for such commercial application for this rather different transportation concept. It has taken the aviation industry half a century to develop this, so perhaps it will take 2 or 3 decades for this to mature for specialized marine craft such as large WIG. The facilities for large CAT and trimarans have developed over a much shorter period—just 2 decades—so there is a precedent.

Safety and Manoeuvrability

Although the stability of WIG flying in strong SEZ ($H/C \sim 0.1$) is satisfactory, it will deteriorate when flying higher into the weak SEZ, particularly if it has a large trim angle. In addition, due to occasional neglect by pilots to adjust the propulsion engine throttle correctly, or when experiencing a strong head wind gust, a craft can fly up out of normal SEZ, into the weak SEZ and experience unstable flight. WIG craft have experienced instability problems and sometimes even craft capsize due to this. Such a terrible event happened to the KM.

PARWIG with rectangular main wing and without composite wing (an outer wing with significant dihedral to provide roll stabilization) have poor transverse stability in case of flying in the weak SEZ, so pilots have had problems when trying to turn or manoeuvre tightly, experiencing unstable coupled transverse and longitudinal motion which are difficult to recover from other than just continuing in a wider curve so as to run flatter. In short, the weakness of some existing WIG is operating in a weak stability zone, in which the pilot has to maintain manual control to ensure that the WIG stays either in strong SEZ or flying out of SEZ, if necessary.

In addition, since the SEZ is rather small, it is a challenge to design a WIG with an automated control system with enough precision to handle the longitudinal and transverse stability requirements of the craft without human intervention. The challenge is therefore that WIG craft of this type (the classic WIG is much more forgiving, but is slower of course) will tire the operating crew quite quickly unless automated flight height control can be developed that perhaps learns from craft and crew responses in real time. While the computer technology is now available in the twenty-first century, adjustment to the delicate requirements of a WIG is still a significant investment of time and money, and will require actual craft to complete the development.

External Noise

The noise level is relatively high currently for WIG due to the high power density propellers. The speed and cost-effectiveness advantage of medium size WIG is suitable for transport to some island vacation resorts located for example around the Caribbean Sea; however, unfortunately, high noise level destroys the quiet and peaceful circumstances and so is a deterrent against opening a fast passenger route in many such situations. Ducted air propulsion has advanced significantly on the last decade based on ACV developments which also had noise problems in the early days, so there is potentially a leverage to be gained by medium size WIG in the near future if the applications can be followed up by WIG designers.

Research and Development—the Seabus-Hydaer Programme

In the late 1990s, a programme to develop a hybrid craft—a hydrofoil supported WIG craft was carried out as a research programme supported by the European Union.

The objective of the R&D programme was to develop a craft that could carry 800 passengers and 120 cars on routes up to 500 nautical miles at a cruise velocity of up to 125 knots targeting routes in the Mediterranean sea. The project was lead by Intermarine, an Italian shipyard, with a total of 11 partners including several universities, Supramar, and the Dutch National Aerospace Laboratory (NLR). The project was completed over 1998–2002.

To achieve service speed in excess of 100 knots the vessel had to be aerodynamically supported, while to minimize propulsion power gas turbine driven water jets were chosen, using a snorkel inlet system similar to that proven for the Boeing Jetfoil (see Chap. 5).

a

b

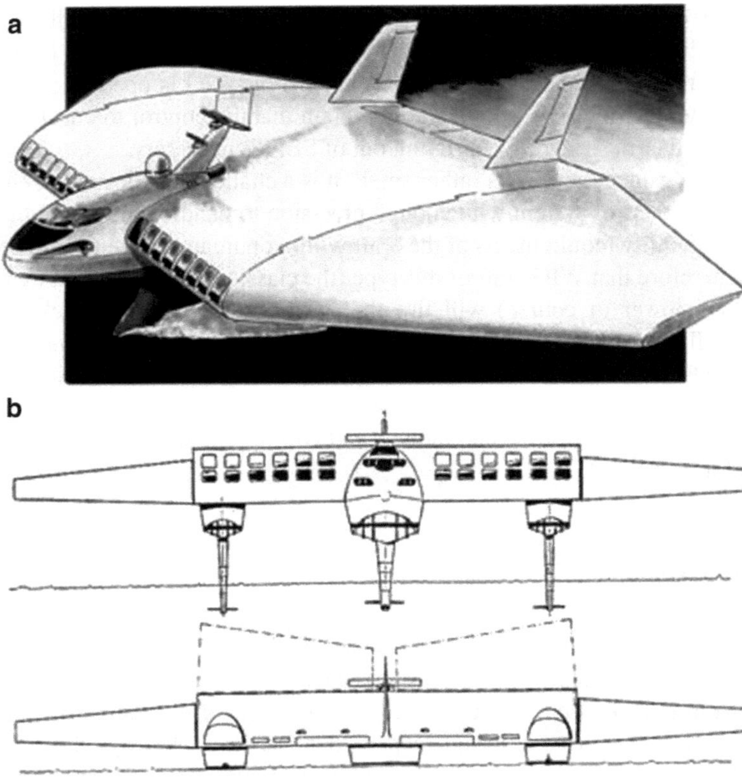

Fig. 3.30 Seabus initial configuration. (**a**) Artist impression in flight, (**b**) above front view in flight, below front view floating, showing possible outer wing retraction for vessel docking

The initial craft configuration had trimaran hulls and connecting hull structure with lifting body shape. Overall dimensions were wingspan 110 m and hull length 53 m. The mean flying height under the wings was targeted at 11 m. Outside of the central body the outer wings were designed to be folded back over the roof for docking and passenger embarking. The hydrofoil struts were also designed to hinge upwards for navigation in shallow water. Main structures were carbon fibre reinforced plastic skins over Nomex foam core material. The water jet intake fins were analysed for impact with large submerged objects [3-9] showing that additional reinforcement with titanium inner skin to the CRP structure would enhance survivability.

The Seabus had flaps and ailerons at the trailing edge of the wings, but it was intended that the fully submerged hydrofoils would be used to control flight elevation and also the submergence of the water jet intakes, see Fig. 3.30.

Airfoil analysis by NLR assisted optimisation of the airfoil geometry [3-10], to deliver the required lift at cruise speed, initially at an angle of attack of 0°. The initial configuration had a relatively high minimum airborne speed of 77 knots, which clearly gave challenges to the hydrodynamic components of the system since fully cavitating geometries for intakes and foils are necessary unless this can be suppressed.

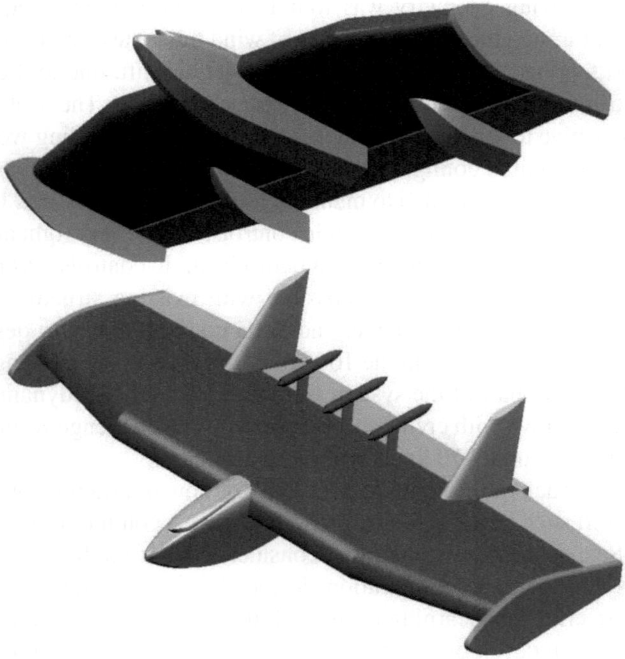

Fig. 3.31 Revised Seabus configuration

Supramar actually found a method which they have patented [3-11], but there are still significant challenges with snorkel type water jet propulsor design for speeds in excess of about 60 knots. This initial configuration also faced the paradox of minimising the forces in the hydrofoil system to minimize drag, whilst needing sufficient force on the foils to control the flight elevation of the craft itself. Supramar's work, resulted in use of air feed grooves at the trailing edge of the main "takeoff" foil to stabilise super-cavitation and air feed grooves placed near the leading edge of the control foil and the vertical struts. The leading edge groove created a two phase boundary layer which is relatively stable at all speeds and normal angles of attack so creating stable control forces at the flying speed of the Seabus. This result is most important for high speed hydrofoils and perhaps also for high speed water jet impellors and inlets. The forces required by the Seabus were nevertheless very high.

During the third year of the project the team agreed to revise the craft configuration to an air propelled WIG form, see Fig. 3.31. This frozen configuration has a central fuselage/hull, large outer side buoys at the main wing tips which were a single contiguous structure, and two smaller sponsons under the wings below the location of the two main fins. The wing neutral angle of attack was raised from 0 to 5°.

The foils were retained for directional control purposes in the second phase. This allowed analysis of cross wind condition operation. The hydrofoils would keep the vessel on track without sideslip, but aerodynamic drag forces were increased, decreasing the efficiency of the vessel in such conditions. The overall aerodynamic efficiency of the wing geometry was found to be approximately equivalent to a freight aircraft through the extensive series of wind tunnel tests carried out by NLR. The wing loading had to be much lower than a jet aircraft, due to the low service speed, and so this overall efficiency was quite an achievement. The Seabus was found to be quite sensitive to wind gusts though, as the light wing loading would suggest.

Just as with the initial configuration, flight control of the craft to ensure the control foils stay in the water needed to maintain flight elevation within a 1-m band and bank angles <2°, through the aerodynamic controls. This is very tight and suggested that the combination of aerodynamic and hydrodynamic controls was not optimal.

The project confirmed that it is possible with existing structural technology, power plant in the 20–30 MW range, and WIG configurations, to design a vessel that can deliver a performance in the 100 knots class for mixed passenger and vehicles. If the hydrofoil part of the system were removed, the aerodynamic properties are at least competitive with civil transport aircraft. The challenge would remain the takeoff speed at just under 70 knots without power assist.

This takes us back to the challenge for WIG of how to arrange optimum takeoff assistance. Craft such as the Swan from MARIC have a configuration not dissimilar to the final Seabus arrangement, if one considers how it might be scaled up. The difference is the bow fan configuration. Seabus has been a tantalizing glimpse at possible future craft. The learning for hydrofoils as a component should have input to hydrofoil design development, while next generation dynamic assist is the challenge before large WIG craft can become a commercial competitor.

WIG Evolution—Current Smaller Craft Development Programmes

We have summarized the main design variations for WIG craft and their classification above. The historical sequence of development actually started with the Russian craft targeted at a mission for logistics at speeds as close to an airplane as possible. The PARWIG craft they developed were technically very successful, and with funding at the military level available were able to demonstrate their usefulness for some years in the 1980s. It was the break-up of the Soviet Union and so the change in mission requirements for the military forces in Russia that stopped their continued development, so rather like the Concorde airliner, technology moved ahead of social need and had to be curtailed. The development of commercial alternative missions did not happen partly because the hydrofoil was so successful in the Soviet Union for river transport between the cities strung along the Volga and Don. The relatively small DACC Volga 2 found some applications but did not make a significant impact on the passenger transport market.

Outside the Soviet Union most western nations were committed to programmes of development for hydrofoils, ACV, and SES in this period, both for military purposes, and commercial applications. Germany sponsored significant development of WIG at a small scale, but the resulting craft did not show enough advantages over

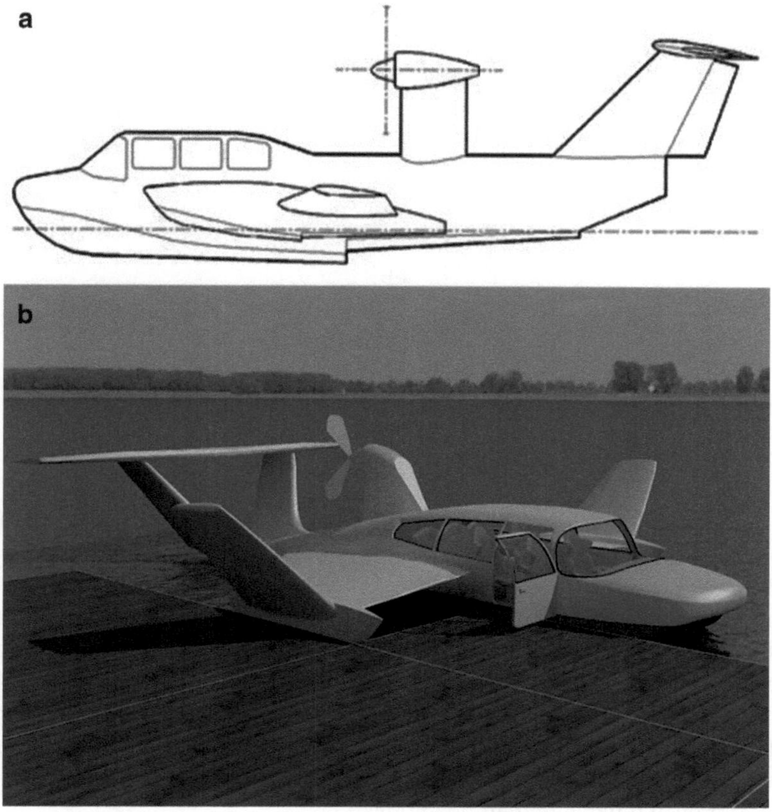

Fig. 3.32 (**a**) Sea Eagle SE-6 WIG craft. (**b**) Sea Eagle SE 7 impression

aircraft or helicopters for the military missions to be adopted for active service. The main challenge was, and remains the sea state limitations for takeoff and landing, and the consequent risks such a craft takes when making an open ocean journey. There are significant areas of the world, both ocean and coastal where this risk can be managed, but at present this does not match with military missions or commercial requirements.

Further development of the DACWIG concept or something similar to it that can lift the sea state for takeoff will be necessary for the commercial attractiveness that can win operators. This requires a scaling up closer to the size of KM, similar to the journey travelled by the catamaran concept, which we explore in Chap. 6.

This does not mean that WIG craft cannot be developed for commercial or indeed utility use. There have been several craft developed in the last decade which are suitable for small scale passenger transport and can operate in coastal conditions. The Sea Eagle SE-6 developed in Australia and Singapore is one of these, see Fig. 3.32a. It is a "classic" WIG so is very simple, with a single pylon mounted 300 shp Lycoming IO-540 light aircraft engine driving an aircraft propeller, the

Fig. 3.33 Iran Bavar 2 patrol WIG craft

prototype is certified and in operation. Sea Eagle is developing a 20-seat craft and has been working on a 7-seat upgrade of the SE-6, shown in artist impression in Fig. 3.32b.

Since 2006 Iran has also been building small military WIG. In 2010, the Navy in Iran showed a squadron of 12 small WIG craft similar to the original Lippisch configuration for coastal patrol in the Persian Gulf, based at the Bandar Abbas Naval Station, see Fig. 3.33. While small at two seats, these craft are armed and are used as a deterrent to illegal marine traffic in their national waters. The craft is called the Bavar 2. While a small craft, it will encourage further paramilitary use of this technology if proven reliable in their operations.

Since 2005 the South Korean Government have been supporting a substantial technology development plan for WIG over the period to 2019. The eventual target is craft with more than 100 t payload and service speeds in the range of 200 kph (110 knots). Shipbuilders Samsung and STX have taken part in some R&D so far, while smaller start-up companies have begun design and production of craft in the 6–20 seat range. Two of the new companies are C&S AMT Ltd who currently produce the Aron-7 craft in series production and have their prototype of the 12-seat Searider on development trials, and Wingship Technology working on a 40-passenger craft the WSH-500.

The Aron-7 5-seat WIG craft from C&S AMT Ltd, Fig. 3.34, complies with IMO rules for operation under marine regulations (classification as WIG Type B). Top speed is 200 km/h with cruise at 120–150 kph, ability to climb to height up to 150 m, with a range of up to 800 km powered by its single Lycoming XP-360 light aircraft engine of 170 shp. A number of the craft are in operation with Korean services, and for taxi service to a luxury resort. Figure 3.34b shows C&S AMT's production shop with one Aron-7 ready for delivery and the prototype 12-seat craft in foreground.

Wingship Technology is working on a size bigger than C&S, with the WSH-500 design for 40 passengers, and the WSH-1500 for 150 passengers. Both craft are planned for gas turbine propulsion running on diesel fuel. An artist's impression of

Fig. 3.34 (**a**) Aron-7 5-seat WIG craft. (**b**) Production at C&S AMT

the WSH500 is shown in Fig. 3.35. A summary of the data is given in Table 3.2 below. The designs incorporate learning from the hoverwing developed in Germany in the 1990s at Fischer Flugmekanik, employing an air cushion under the central hull, tapered anhedral wing form, and twin pusher propulsion. With exception of the cushion support, this overall design has already been proven with the Flightship FS-8 and so has a high chance of success, as long as the economics is manageable. Wingship's first project is to design and build the 40-seater craft, under a contract to South Jeolla Province and Jeju Island and deliver it in 2012 when Expo 2012 Yeosu Korea will be held.

The Aron-7 has shown that practical craft are able to be delivered in Korea, and operational experience is now being built. The Aron has a simple stepped hull and gains its performance through its carbon fibre reinforced resin construction giving it light structural weight. The Wingship's design is following a similar course of using much technology from modern light aircraft for structures and power plant

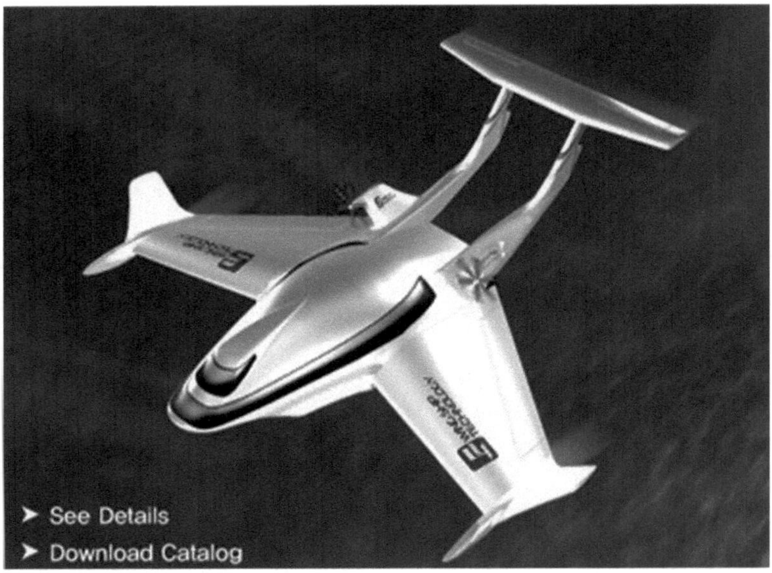

Fig. 3.35 Wingship's WSH-500 passenger craft

while also taking what has been learnt from the earlier technology programmes in Germany, Russia, and China.

In the USA, Universal Hovercraft is continuing to develop personal WIG craft. In the last decade or so, they have refined their UH-18 and 18 hoverwing models which are basically a hovercraft with added on rectangular canvas wings, and extended tail control surfaces. This craft has been popular with home builders for the experience of WIG flight for brief periods rather than sustained journeys in flight. Universal has now turned their attention to the tandem wing concept developed by Jörg in Germany in the 1990s with their UH-TW-1 tandem wing WIG (Fig. 3.36). This design has a relatively modest 35 shp engine and can run at up to 90 kph with it driver aboard. Its maximum flight height is 0.6 m; loa is 5.8, boa 2.4 m, and operating weight 300 kg.

WIG development is still active, and with government support such as that available in South Korea, one may expect progress to be made in the next decade towards medium sized vehicles that are economic. The current choice of light aircraft engines from Lycoming is a conservative approach to start, while it will be interesting to see if turbines running on diesel prove the power plant of choice rather than using automobile turbo diesels in the longer term. Perhaps just as for ACV's it is needed to home in on the optimum overall configuration first. This is still an open question for WIG in the 150–300 kph speed range. There are strong proponents of the different configurations using PAR, blown air cushion, and enclosed air cushion for takeoff that can support efficient larger craft. An arrangement leading to minimised complexity of the mechanical systems would appear to be the best approach.

Table 3.2 Recent WIG craft developments

	ARON-7	Sea rider	WSH-500	WSH-1500	Sea eagle	UH TW-1
Country	S Korea	S Korea	S Korea	S Korea	Australia	USA
Type of craft	Classic	Classic	Classic/air cushion	Classic/air cushion	Classic	Tandem wing
Mission	Utility	Ferry	Ferry	Ferry	Utility	Fun
Status	Production	Prototype	Design	Design	Prototype	Prototype
Max Takeoff weight, t	1.4	3.9	n/a	n/a	2.0	300
Payload	4+1 persons	12+1 persons	40+2 persons	150+2	5+1	1+1
Length, m	10	14	28.5	43	12.5	5.8
Span, m	12	14	27	43	11.5	2.4
Height, m	2.9	3.4	6.7	10	3.5	2.3
Power plant engines,	Lycoming XP-360	Lycoming	2×turbines	2×turbines	Lycoming IO-540	Briggs & Stratton
Power, shp	170	260	n/a	n/a	300	35
Speed, kph max	200	220	175	200	150	90
Cruising	120–150	150			130	70
Range, km	800	800	300	400	250	40
Wave height, TO	1.5 m	2.0	n/a	n/a	0.5	0.1
and cruising	2.0 m	3.0			1.0	0.3
Flying height, m	1–5	2–8	1–5	1–5	1–8	0.3–0.6
Jump altitude, m	150	150	n/a	n/a	50	
Structure	CFRP	CFRP	Al alloy	Al alloy	GRP	Wood/GRP

Fig. 3.36 Universal UH-TW-1 tandem wing WIG

WIG Potential

The challenges limiting the development of WIG for commercial applications are gradually being eliminated as the technology and legislative control of WIG through the IMO develops. The unsatisfactory stability in the weak SEZ has been eliminated in some current WIG craft, which can be flown at all altitudes even with large trim angle.

WIG craft do have significant potential, which still can be used in niche applications where ordinary airplane and boats, ships, or vehicles cannot access. The missions are not dissimilar from other HPMV, such as passenger-cargo transport, salvage, anti-smuggler, border protection, coastal patrol, customs, and some military applications. Let us look at the aspects that may be used to advantage in these applications.

- High speed
  Compared with conventional ship, WIG service speed can be very high, perhaps 200–250 kph, almost ten times as fast as traditional ferries, and also 2–3 times other HPMV, as well as close to the speed of conventional small airplanes.

- Low fuel consumption giving high range
  Thanks to surface effect, the WIG can have high aerodynamic efficiency, i.e. high lift/drag ratio. Its speed is 2–3 times higher than ACV and hydrofoil with the same lift/drag ratio, giving with higher range and lower specific fuel consumption. This can be as low as 20–40 g of fuel consumption per passenger and each kilometre (0.020–0.040 kg/passenger.kilometre) for a WIG passenger craft at a speed of 150–250 kph, compared with ACV and/or hydrofoil with fuel consumption of 0.040–0.060 kg/passenger.kilometre at a speed of 80–100 kph, depending on the power plant employed.

- Cost-effectiveness
  Due to its high speed our comparison is against propeller driven aircraft and helicopters that may be considered as alternatives for the mission requirement. Thanks to surface effect, the installed power of WIG should be lower than that on a conventional airplane, and also much lower than a helicopter. The aerodynamic efficiency of a helicopter is lower as the horizontal rotor generates the lift to support the helicopter flying in air, but not with aid of high speed oncoming airflow. The fuel consumption of a helicopter is as high as 70–100 g/ each passenger.kilometre and in addition the operating and maintenance cost for helicopter is much higher than that of a WIG. The WIG craft can be designed and constructed as ship, not as airplane with facilities for cabin pressurization for example. That means all of design and construction rules and regulations can be similar to that for ship/boat constructions.
- Safety
  Due to its low flying height, about 2–10 m in the SEZ, a WIG always has fairly stiff vertical and transverse/longitudinal stability. In case of an operational failure of engines (collision with flying birds etc.) or other machinery and loss of speed or manoeuvrability, a WIG is able to touch down safely on the water surface and plane to a halt. Currently, airliners can survive such a landing sufficient for passenger escape, but not survival of the aircraft itself. Such situations have been tested during operation of WIG and proven the survivability of machine as well as personnel.
- Seaworthiness
  Thanks to flying in SEZ, and clear from the water surface, the wave disturbance to the WIG is low, and so does not significantly reduce speed and cause motion of craft running over waves. The takeoff ability for the WIG in waves is still the main challenge—once flying the journey is smooth.
- Scaling up
  Since WIG are able to touch down and take off on the sea and take advantage of amphibious ability to land and offload direct at a terminal rather than at a quay, the sea and river surface can be a natural blue express way for the operation of WIG, avoiding major investment in landing runways etc. This offers a very important opportunity for modern transportation for large scale high value freight in the medium future. At present, this travels by air, and has to negotiate the land route to an airport hub, at each end of its journey. Both Boeing 747 and Airbus A340 are unique aircraft that have been operating for nearly 20 years, weighing about 400 t, with a capacity of 400–500 passengers or the equivalent in freight. After development lasting over a decade, Airbus has now begun commercial deliveries of the A380 Airbus with accommodation of up to 800 passengers. There used to be the expectation that industry would eventually create an even larger airplane, weighing a thousand tons, and accommodating thousands of passengers, with various luxury outfit and facilities, such as restaurant, bedrooms, etc. and provide a service closer to that on a ferry ship at higher quality. Such an ideal seems unlikely to be realized as one of the difficulties is the development

of airport facilities for such a large aircraft as has been experienced with the A380. Weight increases with of the cube of dimensional size of aircraft ($W = \text{constant} \times L^{3}$) while lift increases with L^2 and so the runway length has to be extended considerably as aircraft size is scaled up. Currently, only major intercontinental hub airports can handle the A380 aircraft. A WIG, on the other hand, is able to use the sea surface as a natural runway, on which to take off and touch down. It is possible therefore, looking back to the days of large float planes, to take advantage of protected coastal waters near to many major cities as a blue expressway to develop passenger and cargo water transportation with the WIG. The limitations on WIG size are therefore controlled by lightweight structural design, and operating economy which is in itself mainly controlled by power efficiency and fuel consumption.

Currently, WIG craft are normally experimental machines. The most likely service operations that may eventually become attractive enough to support series production are as follows:

- For quasi military applications, such as border protection, offshore patrol, coast guard, anti-smuggling operation, etc.
- Utility operations over swamp areas, ice/snow, on which conventional vehicles and ships are unable to access, for passenger and special personal transportation, as well as for medical and salvage rescue applications.
- Short range passenger shuttle transportation between coastal cities, islands, boarding and landing passengers at beachside terminals or floating pontoons as have been used for float planes.
- For military or commercial applications, and if a large WIG can eventually be developed, perhaps the transoceanic transport mission as proposed by Boeing.

Chapter 4
High-Speed Monohull Craft

From Oars to Sail

We begin our consideration of high-performance monohull craft in the time of the Greek golden age. In the period from 500BC to Roman times, the Greeks were masters of the Mediterranean based mainly on their design of fast man-powered craft driven by banks of oars. The power system advanced from single to double banks of oars and eventually to multi-decked triremes with three banks of oars. The vessels became bigger as the number of oars was increased while the top speed stayed relatively constant at 4–6 knots. Longer and faster journeys were possible as the vessels increased in size and their sea-keeping improved [4-1,4-2]. The addition of sails was a blessing to the motive power—the oarsmen—where the voyage took the direction of the wind. Figure 4.1 shows a reconstruction of a Greek trireme "Olympias" built in the 1980s by the Trireme Trust (see sources) to investigate the performance of the trireme. It is now on display at the Royal Hellenic Navy Museum in Piraeus. Figure 4.2 shows a model of a Greek Trireme in the Maritime Museum in Munich.

The hull shapes used for these ships were relatively slender, the warships having a pointed prow that was at or below the waterline. This improved the hydrodynamic performance as well as being a weapon of war. The main challenge for these craft was the design of the spacing of the oars on each side, both horizontally and vertically. Too close and the oarsmen would clash all the time! The bigger vessels were also a challenge to crew since strong and fit men in significant numbers were required and they had to be able to stay fit to deliver the required power. Food and water on board were therefore as important a consideration as fuel consumption is for modern fast craft. Biremes and triremes continued to be used for commerce in and around the Mediterranean right up until the middle ages. Being equipped with square sails rather than lateen, their directional ability relative to the wind was limited, so it was not until sail arrangements allowing beating up wind were available that

L. Yun and A. Bliault, *High Performance Marine Vessels*,
DOI 10.1007/978-1-4614-0869-7_4,

Fig. 4.1 Greek trireme "Olympias"

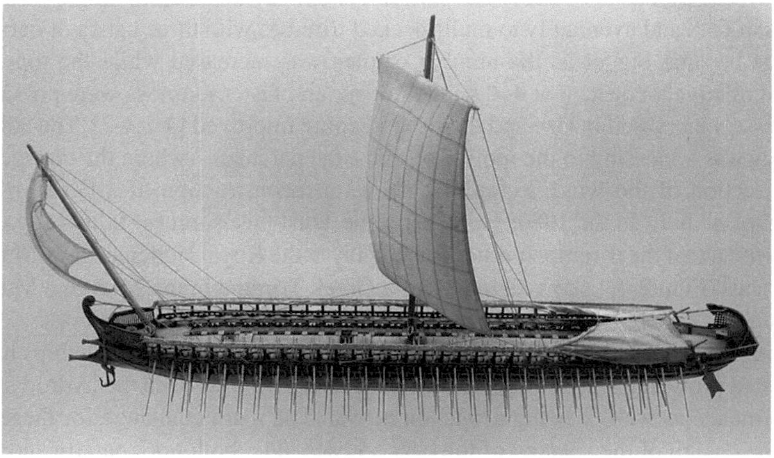

Fig. 4.2 Greek trireme model

sailing ships were able to displace the rowed craft [4-3]. Figure 4.3 below shows a model of a galley built for the Knights Hospitalers in the sixteenth century. Galleys were used for fast transport in Europe in shallow coastal areas up until the seventeenth century as they were not dependent on wind as were pure sailing ships. Some navies, for example, the Russian and Swedish, had galleys through the eighteenth century when the last battles were fought out in the Baltic with craft of this type. The galleys were shallow draft ships and were used in the series of battles between

Fig. 4.3 Knights Hospitalers' galley

Russia and Sweden in the period from 1721 to 1790 when the two nations were locked in a competition for supremacy of the Baltic area. In the long term, both sides showed they were better at defence than attack and so eventually cooperated together against Prussia and other potential threats. The battles were typically fought in line with around 20 ships.

The development of lateen sails and more sophisticated square sail rigs on multiple masts in the Middle Ages was a spur to navigation in the outer oceans; the Atlantic and the Indian Ocean, for Western European craft at least. The more sail was carried, the more ballast was required in the bottom of the hull to keep the craft from capsize if the wind should come from too far on the beam. With open ocean winds, ships could reach speeds in the range 12–15 knots. Ballasting the hull to carry more sail leads to ship hulls with a fuller form, resulting in slower ships rather than faster, even if they could be scaled up to carry more cargo or guns. There was a tendency at that time also to have "castles" at the stern and the bow to allow the officers a safer vantage point when it came to fighting another ship, which was essentially carried out by approaching and grappling with the other vessel.

The English changed this direction of development with lower built vessels and finer bow entry so that the ships were easier to manoeuvre and able to beat further up into wind when setting the sails at an angle. This enabled the fleet to approach an enemy further apart and strafe with a broadside of gunfire maybe avoiding the grapple and fight man to man. The benefits of speed and agility paid off for some centuries for the British from military point of view [4-4,4-5] as their horizons moved from the Atlantic into the Indian Ocean and Pacific.

Fig. 4.4 Clipper ship

While competition was strong between the Portuguese, Spanish, French, British and Dutch in this arena, the British were most often able to win a confrontation through their superior performance. The experience with military ships in this period was also put to use for commercial craft in the Atlantic as the Caribbean and Americas were colonized. The faster ships could deliver colonists, slaves and basic supplies for construction, the better, as competition between nations was significant through that period! Discovery of precious metals and a growing trade in products such as sugar meant that the return journey was also important.

European exploration and trade expanded from the Atlantic to worldwide in the seventeenth and eighteenth centuries. Before this, it was only the Chinese that had completed their exploration of the globe in the fifteenth century with huge seagoing Junks [4-6]. These craft were seagoing communities, built for endurance not for speed. Unfortunately, internal events in China just after these explorations stopped the development, and the nation became insular for several centuries.

Endurance combined with the need for significant cargo volume to return a profit meant that commercial sailing ships needed to increase in size for trade from Europe to the Far East and to develop Australia which had been reached in the seventeenth century by Europeans. This leads to development of the Clipper form [4-3] with sail plans designed for sailing closer to the wind. Clippers literally raced between Europe and India and the Far East, as the first to arrive back with their cargo of tea or spices would get the highest price on the market in Amsterdam, London, or Boston, see Fig. 4.4. They were also used to take emigrants out from England to Australia and return with cargoes of wool.

The fastest clippers were able to reach 18–19 knots in open ocean winds and were entering the speed range where drag from the waves generated by the hull movement through the water was a significant force. Clippers survived in commercial service until the late nineteenth century as commercial work horses [4-7].

From Sail to Steam

We have touched on the transition from sail to steam in Chap. 1. Initially, mechanical power was applied to relatively small craft such as yachts and harbour tugs, followed by gun boats for a number of navies, such as Germany, Britain and the USA. The success of gunboats was then repeated at larger scale with steam power being installed on warship designs that still had a full set of sails for offshore voyaging point to point; the British HMS Warrior is an example of this, Fig. 1.4. Steam turbines scaled up the power available so that by early twentieth century very large warships could be produced—the most famous being the dreadnaught battleships and the equivalent German designs. The advent of the heavily armed and armoured battleship was also a spur to development of the fast attack craft—the torpedo boat able to approach so fast that the battleship could not react, and then launch a weapon at the softest part of the battleship to sink it [4-8,4-9].

Such tactics were possible while a battleship was close to the coast or its base port, while out in the ocean a different type of fast craft was required. The logic used by Navy commands for steam-powered warships in battle offshore followed the approach in previous centuries where large battleships "of the line" were accompanied and protected by smaller handy craft that could sail much faster when required and draw fire away from the heavy guns—the frigate.

The commercial world also changed as steam took over from sail in the late nineteenth century with the development of the passenger vessel to take emigrants from Europe to the USA. Clearly, the quicker the ship could make the return journey, the more profit could be made by the owner. The high-speed passenger steamship was developed into the Atlantic liners of the twentieth century making regular journeys between Europe and the UK to New York. The ships also competed for the fastest time across the North Atlantic in a race that became known as the "Blue Riband" (see resources internet listing).

These ships were of a relatively fine shape. While remaining a displacement ship at their service speed, their sheer size allowed the speed to be high even across the rough seas of the North Atlantic. While air transportation has taken over in the second half of the twentieth century as the mass transport system for long distance travel, liners have developed into the modern cruise ship. High speed is now used to move from one visiting location to the next during the night hours so that the clientele have maximum time available at the next port or area of special interest.

Returning to more mundane commercial duties, ferries across and along rivers were able to deliver a more reliable service once mechanical power was incorporated, and the progression began through steam reciprocating to steam turbine to diesel and to gas turbine power.

Refining the Hull Shape

Unless the dimensions of a hull are scaled up, over a certain speed the drag of a displacement hull will rapidly increase due to friction drag increasing as the square of speed, and the waves generated by the hull also generating drag forces at a rate proportional to speed. Above a Froude Number of about 0.7, the power necessary to drive a ship faster becomes prohibitive unless the hull form can be changed so that rather than pushing it forward with brute force the forces are used to lift the craft and reduce the "wetted area".

This was the idea that pioneers developed in the mid-nineteenth century and kicked off the development of the plethora of types of fast craft including a number of variations on the single hull geometry. Let us start with the physics, and then come back to the concept development.

Froude Number is a very important coefficient for hydrodynamics, related to the drag, and wave and spray making of the craft moving on the water with respect to the craft displacement. It is defined as:

$$\mathrm{Fr} = v / (g\Delta 1/3)^{0.5} \quad \text{or} \quad v / (gL)^{0.5}$$

where

v = craft speed, m/s
Δ = volume displacement or weight of the craft, m^3
g = acceleration of gravity, m/s^2
L = representative length, either hull waterline length or wetted length for planing craft

A hull moving through the water will have several drag forces acting on it. The two principal forces are from friction and surface wave making. First, there is the friction drag from the immersed surface. This gives a force proportional the area, the roughness of the surface and the speed. Second, the hull geometry will generate waves on the surface from the bow, the fore body at the "shoulder", the aft body at the "hip" and at the stern. Each of these wave patterns will interact and as speed rises will cause a series of "humps" in a curve of drag rising almost linearly. The wavelength of the patterns generated increases with speed until the bow wave and stern wave hollow dominate, and the drag increase is more nonlinear. Since frictional drag also increases with a square law, this defines the practical upper limit of speed for displacement ships in calm water.

Out in the ocean with wind waves, these apply additional drag forces to the hull, slowing it down, and also causing it to pitch, roll and heave. These motions again create additional hull wave making and increase the frictional resistance, with a periodic variation like the waves themselves. Yes, a complex situation that can today be calculated (simulated) on a computer, but until the nineteenth century was not understood, until William Froude and some other Naval Architects developed theories and used models towed in long water tanks to verify the theories [4-10,4-11], publishing their results in learned society journals. The research was then used to develop hull shapes to minimize the wave making at the desired service speed and to accurately determine the resistance so as to specify the power that was needed to be installed—this had been an art rather than a science until then. If you can predict the drag, you can also develop faster vessels reliably! The next problem is how to reduce the friction and wave making itself.

Skimming Over the Surface

A round flat pebble can be skipped on the water surface so long as it is given a high initial velocity with a suitable planing angle. The necessary pre-requisites for a surface planing object are:

- Flat planing bottom
- Proper trim angle, i.e. the angle between the planing bottom and velocity vector of the object
- High enough speed

The Frl at which lift forces become significant is around 0.7, and above 1.0 true planing is possible.

Taking a flat plate as the planing object, the hydrodynamic lift of the plate can be written as:

$$L = \frac{1}{2}\rho v^2 S C_L$$

where

L = lift
v = speed
S = area of the planing surface
C_L = lift coefficient

The lift coefficient C_L will be a function of length/beam ratio, the planing surface area, trim angle, etc. It may be noted that at low speed, a planing surface will develop a lift force, as well as drag, but not sufficient on its own to support the whole weight. The water volume displaced by the hull will therefore be less than the displacement at rest, but the hull will not have risen out of the displacement "cavity".

A high-speed planing hull is similar to the pebble, as shown in Fig. 4.5, which shows a craft without steps, and Fig. 4.6 showing a craft with transverse steps, the function of which will be explained later.

From Fig. 4.6, it can be seen that the forward transverse section is sharp for parting the water before skimming, or take off, and the mid as well as stern sections are wide and flat with a small dead-rise angle β, which fits with the requirement for skimming.

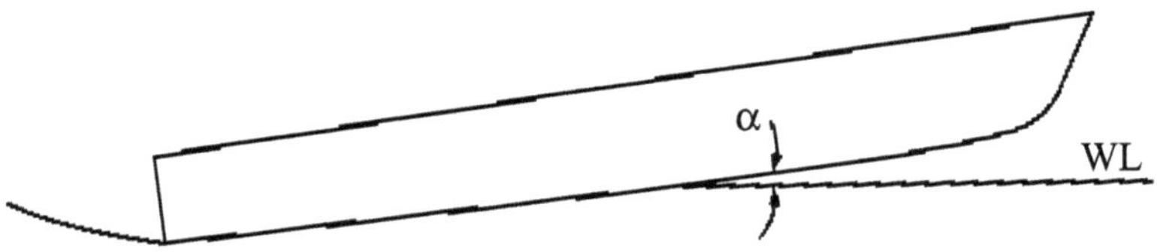

Fig. 4.5 Planing hull without steps

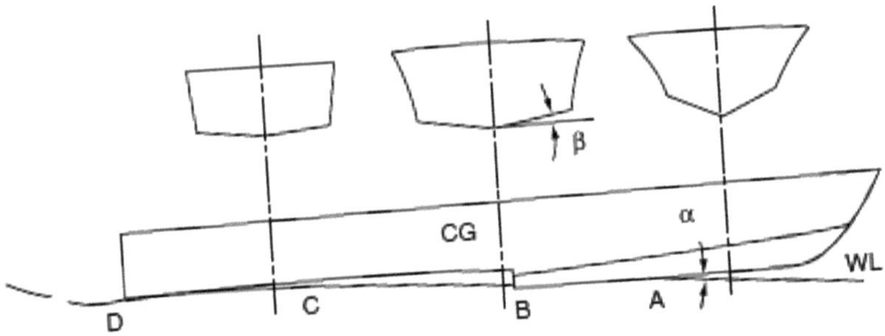

Fig. 4.6 Planing hull with steps

When the craft moves at slow speed in displacement mode as an ordinary ship, this will be accompanied with high wave making as shown in Fig. 4.7a, where the greater part of the craft weight is supported by static buoyancy. With increasing speed, the hydrodynamic lift gradually grows, thus lifting the craft up and reducing the cavity depression as shown in Fig. 4.7b. In this mode, wave making is broken down and replaced by generation of much more water spray from the bow, and a flatter stern wave form. As the craft accelerates further, the weight will be supported by hydrodynamic lift from a reducing surface area, thus the whole craft will rise up and only a small part of the hull will interact with the water surface, see for example Fig. 4.9 The wetted surface decrease as speed increases means that drag forces will increase more slowly for a carefully trimmed boat. The trim angle may also increase due to the hydrodynamic lift acting on the fore part of the craft, and so trimming flaps or other devices are necessary at the stern to control craft attitude at high speed.

Planing surfaces are more efficient when they are wider and flatter; when they have high wetted beam/length ratio (B/L, where L is wetted length, and B wetted beam of the craft skimming on the water surface); and when the dead-rise angle is small. As speed increases, hydrodynamic lift increases, and so the wetted surface, draft and trimming angle all reduce and the centre of lift moves to the rear part of the craft bottom. The craft will run smoothly and comfortably with little spray at this high speed so long as it has longitudinal balance.

From Fig. 4.8a, b, one can see the several longitudinal rails along the bottom of the craft, used to control spray generation and ejection. The water spray can be suppressed and water flow improved by use of guide rails a little like fences installed on airplane wings to suppress transverse flow. The rails improve the craft lift, reduce external spray and make it more comfortable for the crews as well as passengers on the craft.

Fig. 4.7 (**a**) Craft at slow speed. (**b**) Craft planing

Stepped Hulls

A fast planing craft can be supported on the water by a small wetted surface and trim angle, so it can be useful to introduce one or more transverse "steps" on the bottom of the craft as shown in Fig. 4.6. The step is generally located at the middle of the hull bottom, so that the craft is supported both by the forward wetted surface (AB) and rear wetted surface (CD), thus decreasing the total wetted surface and resistance. This is because the hull surface at the mid part of bottom is clear from the water.

More than one step can be introduced into the planing surfaces of a fast boat, as shown in Fig. 4.9. Two, three and even five such steps have been used in different

Fig. 4.8 (**a**) Spray rails. (**b**) Spray rails deep Vee craft

Fig. 4.9 Stepped hull craft at high speed

designs. One benefit that the step introduces is that the angle of the planing surface ahead of it can be optimized to give smoother take-off into planing. The use of more than one step also allows the hull static trim to be set up for the planing condition separately from the displacement condition.

The vertical or near-vertical rear facing wall of the step also has the special function of entraining a controlled flow of air behind it. If the step is smaller close to the chine or outer extent of the hull bottom, it will suck in less air and operate more on vapour generation, disturbing the flow past the hull less at this point [1-5,4-12].

The combination of stepped hulls with the deep V form offers a powerful combination for the offshore powerboat. We will discuss that below. Firstly though, we take a step in the opposite direction.

Flat-Bottomed Boats: The Air Boat

Not everybody needs to go fast out in choppy conditions. In Florida, there are the everglades, a vast area of swamp lands varying between shallow open lakes of water, river like waterways, areas of water that are covered with long grass and mangrove forests. The flora and fauna are a big attraction for nature lovers and hunters. V form and stepped hull fast boats would churn up the shallows. A water screw would also churn up the bottom. The water depth or lack of it means that wave height is very small. Basically, what fits this environment is a shallow punt like hull. Indeed, pole-driven punts are used by hunters for stealth.

If you want to go fast in this environment, the obvious propulsion is an air propeller (as long as you don't mind the noise). That is essentially what an air boat is (Fig. 4.10). It literally slides across the water surface.

Fig. 4.10 Air boat sliding around corner

Apart from care to ensure that the longitudinal centre of gravity is close to but just behind the centre of area of the plan form, there is not much finesse about an air boat. They can also be agile, since with no water-piercing appendages, it is possible to set an airboat sliding sideways and use thrust to take it around the corner, in a similar way to an amphibious hovercraft, even performing a pirouette as small ACVs often do. The crafts are not recommended out in open waters though.

The Deep V Hull Configuration

A fast boat for racing offshore needs a different approach to an air boat. We have introduced the planing craft with low dead-rise planing surfaces canted upwards from the keel line. In calm waters, such a craft can be fast and reasonably comfortable, though unless the bottom surface is carefully shaped with propeller thrust line just right, it is possible that unsteady forces may develop at speed and the boat perform a porpoising motion. Addition of steps into the planing surfaces can solve this problem and also improve transition to planing.

Initially, Naval Architects created hull geometries with bow and mid transverse section designed as round bilge and continuing to the stern with a widening flat bottom and a hard chine form at the stern, Fig. 4.11a. As speeds were increased, this form was developed into the low dead-rise vee from hard chine planing craft, Fig. 4.11b. These developments were driven by powerboat racing in the years between the first world war 1919–1939 and subsequently by the military needs in World War II for fast vessels to chase enemy submarines and destroy them with torpedoes while they "breathed" on the surface. Large numbers were built in the USA and UK as described in [4-13,4-14]. This form gave the following characteristics:

1. Minimized slamming force in waves at the bow, due to fine entry at bow and rounded bilge shape transverse section at amidships for the semi-planing craft
2. The hard chine at the stern for higher speed craft gave a good powering performance in waves, and only a small drag penalty in calm water due to full planing action

These craft performed well, but still had a significant speed loss in a seaway. The answer to this was developed in the mid-twentieth century through the challenge of offshore racing. Race boat designers experimented with planing surfaces having a much higher angle, and in addition, a number of longitudinal strakes along the planing surfaces, as well as a sharper bow shape. This shape enabled the boat to slice through waves rather than bounce across them. Include one or more steps into the planing surface design and high-speed stability is further improved. This is a deep Vee hull as shown in figs 4.8, 4.9, and 4.12 below. The boat in Fig. 4.9 also has several steps for improved dynamic trim control. Key features of a deep Vee craft are:

- The bow transverse section is a V formation with large dead-rise angle
- The bow form is rather fine and long, having a curved keel line that ends horizontal in the region of amidships
- The transverse section aft of amidships also retains a significant dead-rise angle (β=10–20°), so as to decrease the wave slamming forces on the flat surfaces

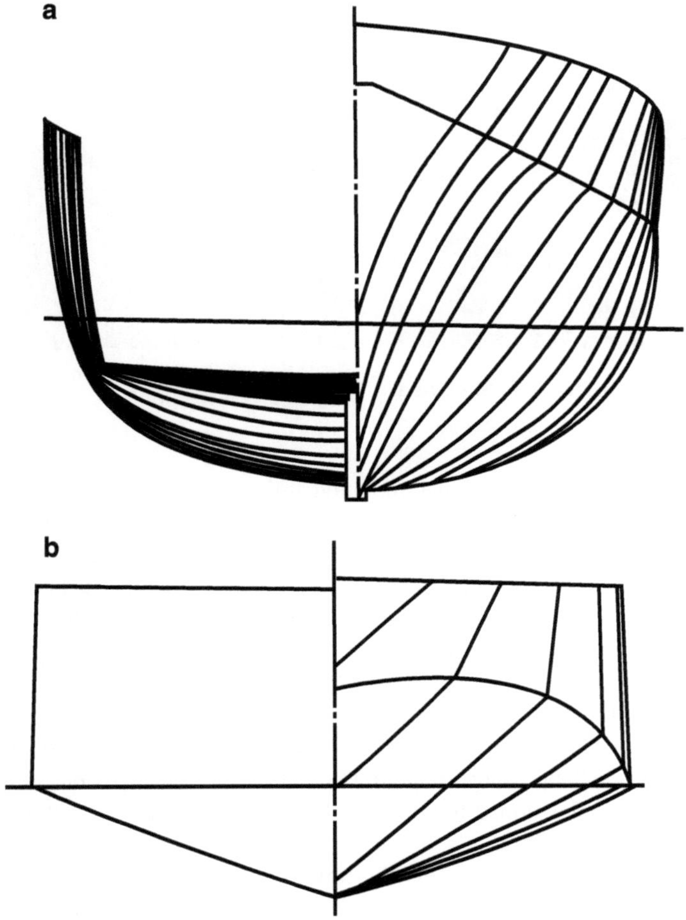

Fig. 4.11 (a) Round bilge. (b) Hard chine planing craft

The advantages of this configuration are:

- Sea-keeping: Due to the greater draft of the Vee transverse section, the dead-rise angle and damping of vertical and longitudinal motions, as well as small wave disturbance force, the craft will have smaller pitching and heaving displacements, vertical acceleration, slamming and smaller probability of bow submergence in waves, and emergence of the propellers at stern as waves pass, compared to round bilge displacement forms.
- Fine manoeuvrability, and course stability.
- Lower speed loss in waves.
- Simple hull structure using single plane curves, giving reduced production cost.

To achieve high speed, a planing craft needs to have an efficient propulsion system. Displacement craft such as Turbinia (Fig. 1.3) had canted shafts under the stern

Fig. 4.12 Hard chine fast ferry Silvia Ana

of the hull driven from the turbines-mounted amidships with propellers mounted on the end (Fig. 1.5). At 30 knots, propellers such as these can deliver thrust efficiently. As speed is increased above 35 knots, the suction pressure on the back of propeller blades reduces below atmospheric, and so water vaporises at the surface and creates a cavity. If this process is unsteady, it can cause severe damage to the blade surface. Special blade geometry is therefore necessary to actually cause a steady cavity—the super-cavitating propeller. Development of these propeller types was a significant focus for shipbuilders and Navy R&D in the mid-twentieth century for fast patrol craft [4-8,1-2]. The problem with this configuration is that the shaft and its intermediate supports add "appendage drag" to the craft. In the same period, water jet propulsion units were developed for small craft operating in shallow waters with flush inlets—no appendage drag.

Initially, it was not realized that the flow regime at the intake to a flush entry water jet at the base of a hull could actually reduce drag, but through development programmes in the US for the SES, and fast military craft, this was demonstrated. This encouraged water jet manufacturers to design units for larger power ratings, leading eventually to the size useful for passenger catamaran ferries, and eventually large monohull ferries.

Due to the simple hull structure and lower investment compared with the more complex HPMV, monohull craft have continued to be popular within their envelope of efficient operation. The availability of large marine gas turbines with improved fuel consumption, and also high-power high-speed diesel engines, it has made it possible to design efficient large deep vee craft for ferry applications. The use of modern high-power water jets for propulsion has further improved their performance and reliability.

Figure 4.12 shows the Deep V type monohull car-passenger "Silvia Ana", with overall length 102 m, beam 15 m, draft 2.4 m, accommodating 148 cars, 550 passengers, powered by four high speed diesels of 5,498 kW each, driving four water jet propulsors, giving a service speed of 40 knots. Shipbuilders in Spain, Italy and France have built a number of fast ferries in this configuration for example.

Racing Craft Development

We have described above the development of the deep vee configuration for offshore power boats. In recent years—late twentieth century and first decade of twenty-first century, offshore racing has developed into a number of sponsored racing series where craft have to negotiate a “closed circuit” loop type course. The course length is long enough to demand high top speeds, while short enough and close enough to shore so that spectators can have an exciting event to watch. We give some internet locations for racing information in the resources at the end of the book.

Racing craft have developed both bottom geometry with multiple steps, upper hull aerodynamics and power systems with surface riding ventilated propellers, see Fig. 4.13. Speeds are well above 120 kph. The accelerations applied to crew are extreme, and small errors in navigation can cause craft to leap from wave tops. The hulls of such craft are now built in carbon fibre impregnated with resin for light weight, stiffness and strength. Endurance for both structure and mechanical systems is defined by the race length, rather than by long-term operation and so both structure and equipment design can be taken to its limit, rather like in motor car racing.

The tendency for propulsion has been towards surface riding super ventilated propellers. While water jet propulsion works well in the speed range 30–70 knots, the intake geometry has to be carefully designed so that the impellor is neither choked or starved over its operating speed range in the craft since either case leads to unstable cavitation and damage to the impellor.

The surface riding propeller system is not as efficient, but is simpler and lighter to incorporate into a racing craft and much easier to replace.

Fig. 4.13 Arnesen surface drive

Fig. 4.14 racing craft open cockpit

Fig. 4.15 Open class offshore racing boat

Offshore powerboat racing is an exciting sport, and one where small errors by the crew, or a wind gust, or breaking wave can upset the boat, particularly in cross wave and down wave directions. Designers are constantly searching for the small design changes that can give a boat resistance to these events and maintain top speed. At such speeds, aerodynamics are very important upwind to prevent pitch-up, as is the re-entry into waves, particularly down-wind with the possibility of submarining. Figure 4.14 shows an open cockpit deep Vee racing boat with stern-mounted Z-drive propulsion, the higher power alternative to an outboard motor where the engine is installed just forward of the transom. With all this weight at the stern, a craft like this benefits from a water ballast system installed forward to control the centre of gravity.

Figure 4.15 shows a gas turbine-powered unlimited Class racing craft with enclosed cockpit and surface riding propulsion. The propellers are aft of the transom with the shafts supported on brackets at that point. Both these racers have systems for moving the propellers themselves to provide steering rather than having a rudder, so minimizing appendage drag.

Super Yachts

Super yachts are a class of HPMV that has developed from the earlier yachts built for cruising in the nineteenth century once mechanical power became available [1-2]. The beginnings of this class of vessel were described in Chap. 1. Speed is an important part of the design of such a yacht, though loitering performance and comfort in a seaway is also very important. Typically, such a yacht will be designed for a top speed in the 20–30 knots range, while being able to make long distance transit at reasonable speed, generally at 12–18 knots. For comfort, such craft are fine at the bow while relatively beamy and will have a displacement or semi-displacement overall hull form. Using a displacement form weight is less of a priority allowing the internal outfit to be along the lines of a high class hotel for the guest passengers. Figure 4.16 shows the Octopus, built for Paul Allen, co-founder of Microsoft.

Variants of this have developed in the last two decades. First, there is the super yacht built with relatively little accommodation, while having space for luxurious entertaining, and often being much faster craft with 30–50 knots as a dash capability. Figure 4.17 shows the Millennium 140, the world's fastest in 2010 with a top speed of 70 knots in calm conditions. Development of carbon fibre hull construction has assisted the development of these craft allowing light structure, open internal design and smaller power plant installations to achieve a given performance. The

Fig. 4.16 Super yacht Octopus

Fig. 4.17 Fastest super yacht Millennium 140

top speed of the Millennium 140 demands prodigious power to be installed; in this case 2 × Paxman 18Vp185 Diesels of 5,300 shp each driving a steerable LIPS Waterjet, and 2 × Lycoming TF-80 Gas Turbines each of 4,700 shp together driving a third LIPS water jet (Lips water jets are part of Wartsila).

A shipyard in the north of Denmark, Danish Yachts of Skagen has built a super yacht "day boat" aiming at higher efficiency from a monohull planing configuration. The Danish Yachts 116 is built in carbon fibre and has a flat vee cross-section to the hull aft of amidships. The curved bow section has spray rails that do not extend right to the stern. The 38 m boat can achieve 50 knots with just two MTU 16V400093L diesels of 3,440 kW each, see Table 4.2. The hull form promised to have very little drag hump under planing speed and no need for stern trimming devices according to the designers, and this proved to be the case in trials.

This hull form can be optimal for a high-speed boat use for recreation, since its use will be primarily in mild sea conditions, so the smooth acceleration and power economy are most attractive. Where a vessel is to be used for ferry purposes, it will encounter heavier sea states in service, and so the aft cross-section is optimized with deeper vee and spray strakes, an example is shown in Fig. 4.18.

Access to the more traditional super yacht is itself an HPMV mission for the larger yachts since they will often moor offshore and need to transport their limited number of guests ashore quickly. This has led to the fast yacht tender with speed in the range 30–50 knots. An example, the Windy W290S is shown in Fig. 4.19. Technical details are listed in Table 4.2.

The real challenge of super yacht design is to balance between supplying high performance for logistics, while also being able to provide a very comfortable accommodation, including entertainment spaces. There are interesting comparisons to draw here between the demands for high-speed ferries and the larger cruise ships, and on the military side, the Corvette or Frigate where in all cases personnel will live on board for at least a night or so, rather than just occupy a seat for minutes or hours.

Fig. 4.18 High-speed monohull ferry Vesuvio Jet 4

Fig. 4.19 Super yacht tender Windy W290S

Design Challenges and Applications

So what are the key characteristics of a monohull HPMV? The following attributes are all important to get right for the prospective mission:

1. *High speed* in both calm water and waves, with minimum speed loss in a seaway. The speed needs to be tuned to the mission, so as to optimize powering, and fuel payload.

2. *The monohull has compact dimensions for a given payload* compared with other HPMV, such as ACV, WIG and catamarans, allowing low hull structural weight and higher manoeuvrability, while payload volume is more limited.
3. *Simple technology*, no complicated outfit or structural configuration such as air cushion system and flexible skirt for ACV, hydrofoil system for HYC, air wing system for WIG. This can become blurred if gas turbines powering is used, or too complex a stabilization system.

   The challenges to designing these craft for high-speed operation in a seaway are:
4. *Stability*: The hull bottom geometry and configuration of steps and spray rails are critical to dynamic stability both in calm water and waves. A hull with incorrect curve to the keel may be unstable in pitch, leading to the porpoising motion. Hulls without spray rails correctly designed can also have tendency to roll back and forth.
5. *Sea-keeping*: when running in waves, monohull craft often rise out of the water and the re-entry into the next wave gives a shock force called slamming due to the high vertical deceleration, and associated additional drag and speed loss. The impact is highest for flat bottom craft and lowest for deep vee configurations. Such impacting force and high vertical acceleration can cause crews and passengers discomfort, while equipment and local hull structure loadings are very high and have to be carefully designed for.
6. *Short range* due to high main engine-installed power and fuel consumption, while the small dimensions of the hull give limited fuel tank space, hence the need for optimisation mentioned above.

These design challenges have been met for fast passenger ferries by using lightweight structural materials for the hull (welded marine aluminium) and compact main machinery (lightweight high-speed diesels and gas turbines). When combined with modern water jet propulsion and medium vee form bottom, it has been possible to scale up the monohull to give high passenger and vehicle payload and operating range suitable for fixed route operations.

Delivery of the comfort levels demanded by passengers is now possible by using stabilizers; either moveable fins just aft of amidships, or by flaps or "interrupter devices" at the stern. These can stabilize roll and pitch motion by dynamic response and optimize trim at varying speeds by varying their neutral position. Development of computer control systems since the mid 1970s has provided increasingly reliable automation to these systems.

Many coastal ports have existing jetties and quays that have been used for conventional displacement ferries for many years. The adoption of a monohull high-speed ferry assists to minimize the upgrade cost for an operator, compared to the specialized docking facilities needed for catamarans or other multihull craft. The attractions of increased work capacity from a catamaran are strong, but the total investment has to be taken into account. The first option is therefore often to look for a monohull that has improved work capacity compared to the existing operation, with minimized terminal investments, and to use the increased income from the initial vessel upgrade to fund terminal upgrades that can prepare for the more demanding HPMV types.

Fig. 4.20 Fast strike craft

Fig. 4.21 Fincantieri coastal patrol craft for Italian coastguard

In the military world, monohulls continue to be refined for applications such as fast strike craft Fig. 4.20, coastal patrol for fishery and piracy control Figs. 4.21–4.23, and for core navy missions that are carried out by Corvettes and Frigates. Apart from strike craft, all these vessels use a composite "semi planing" form with relatively fine bow entry, round bilge fore part, and a flat bottom or small dead-rise form aft with small or no bilge curve. The bow has a marked flare at the upper part to deflect waves in rough sea and provide lift to avoid submarining. This form gives

Fig. 4.22 The latest vessel series in the US coastguard fleet—the "Sentinel" class

Fig. 4.23 Fleet of offshore patrol boats built by Austal for Yemen coastguard

Fig. 4.24 Offshore crew boat

some planing lift at full power while leaving the craft substantially supported by buoyancy. A high L/B ratio restricts drag development at higher speeds, and the fine form gives low-speed loss in higher sea states. Motions are controlled generally by strakes called bilge keels along the mid-line of the bilge curve, and similar to comments above for ferries, stabilizer fins and transom flaps or interrupters can be installed for dynamic control. The same hull form is used for fast crew boats that are used to take workers out to offshore platforms in the Gulf of Mexico in the US, Mexico and Venezuela, as well as at a number of locations in the Far East. An example is shown in Fig. 4.24.

A strike craft with a design speed exceeding 40 knots generally demands gas turbine propulsion, and this in turn limits the range and endurance of such as vessel. Offshore patrol requires a long endurance at lower speeds, as well as a dash capability. Modern high-speed diesels when fitted with turbochargers are now light enough and efficient enough to propel a semi-planing vessel a dash speeds above 30 knots and so deliver against the typical mission requirements for coastal patrol.

The big advantage of this type of craft is that it can be adapted for long period slow speed operation by installing secondary lower power engines, while the hull form, being based on displacement operation, can be designed to give kindly motion at these "loitering" speeds. Essentially, this type remains the premium choice for long endurance missions.

The high-speed planing craft has developed and kept a specific niche in the high-speed water transportation market, while designers and operators have found other craft concepts that have benefits in terms of payload space, stability in a seaway, etc. to expand the market as such.

Tables 4.1 and 4.2 below give the principal dimensions and key features of some example civil monohull craft, while Table 4.3 details example military patrol craft.

Table 4.1 Example high-speed civil monohulls

Name	Catalina express	Capri jet	Ono-ono	90m Fast Ro-Ro Ferry	Eagle express	Corsaire 12,000	Gotlandia II MDV 2,000
Country	USA	UK	Australia	Japan	Australia	France	Italy
Builder	West Point Shipyard Inc.	FBM Babcock Marine Group	Austal	Mitsubishi Heavy Industry	SFB Shipbuilders	Aker Finnyard Group	Fincantieri Centieri Italiani SpA
Delivery	1986	1988	1994	1997	2000	2000	2006
Mission	Passenger ferry	Passenger ferry	Passenger ferry	Passenger + vehicle	Passenger ferry	Passenger + vehicle	Passenger + vehicle
Passenger	149	350	450	423	350	1,000	780
Vehicle				106		210	160
Loa (m)	27.43	41	48.0	101.0	35.0	119.0	122.0
Lwl (m)	26.5	39.5	41.3	90.0	31.7	105	112.2
Beam (m)	6.4	7.8	9.0	14.9	7.0	15.7	16.65
Draught (m)	1.37	1.1	1.2	2.5	1.9	2.6	2.9
Crew	4	5	11	11	6	22	20
Deadweight (t)	n/a	27	n/a	n/a	n/a	500	580
Max speed (kt)	32	33.5	25	42.4	32	36	40
Engines	2 × Detroit diesel 16V149T1 BDDEC	2 MTU 12V396TB63	3 MTU 16V396TE74L	4 MTU 20V1163TB73L	3 MTU 12V2000M70	4 SEMT diesel	4 MAN B&W 20RK280
Propulsion power (kW)	2 × 1,491	2 × 1,180	3 × 1,960	4 × 6,500	3 × 2,100	4 × 8,100	4 × 9,000
Propulsion	2 × Propeller	2 × MJP WJ	3 × WJ	4 × KaMewa WJ	3 × FP prop 5b	4 × KaMewa, 125II WJ	4 × KaMeWa, 140 SII WJ
Structure	FRP sandwich, Airex core and fire retardant resin	Aluminium	Aluminium	Aluminium	Aluminium	Steel hull Al super-structure	Aluminium
Frl/SS	0.85	0.86	0.59	0.69	0.89	0.54/4.5 m	0.59

Table 4.2 Example high-speed civil monohulls

Name	Sofia M, Carlotta M	Kalymnos Dolphin	Octopus	Millenium 140	Windy Dubois SR 52 Blackbird	Shooting star
Design	Air Naval 38	EM20	Lurssen Werft	Mulder Design	Windy Boats AS	Project 116, Espen Øino Aerocruiser 38 II
Country	Italy	Greece	Germany	Holland	Norway	Skagen, Denmark
Operation	Italy	Greece	Mediterranean	Caribbean	Worldwide	Europe
Year	2010	2010	2003	2004	2010	2011
Builder	Air Naval	Epsilon Marine	Lurssen	Hardinxveld, Giessendam	Windy Boats, Skarpnes, Norway	Danish Yachts
Application	Ferry	Ferry	Super yacht	Super yacht	Super yacht tender	Super yacht day cruiser
Passengers	210	124	26	10	11	14
Crew	6	2	57	6	1	2
Cars	–	–	7 tenders, 2 helicopters	Several tenders	–	5 m tender + jet ski
Displacement (t)	242	48	9,932	n/a	12	112
Loa (m) (Lwl)	37.6	19.95	126.2	42.2 (32.75)	16.1	38
Boa (m)	7.0	5.3	21.0	8.25	4.5	7.5
Hull depth (m)	5.0	3.0	12.5	6.0	3.0	7.0
Draft (m)	2.4	1.05	5.66	1.88	0.5	1.5
Speed (knots) Max	35	30	20	70	46	50
Cruise	32	25	17	50	40	25–40
Main engines	3 MTU 12V2000M72	2 × MTU 8V2000M93	8 Mercedes Benz diesels driving 2 ABB electric motors	2 Paxman 18Vp185, 2 Lycoming TF8 GT	3 Volvo Penta IPS 600	2 MTU 16V 4000M92L
Power (kW)	3 × 1,080	2 × 895	8 × 1,780 driving 2 × 6,000	2 × 4,000, 2 × 3,500	3 × 325	2 × 3,440
Propeller	3 × FPP	2 × 4b FPP	2 CPP	3 LIPS WJ	3 × FPP	2 × MJP WJ
Stability fins	2 midships stern interupters	–	–	–	–	–
Structure	Al alloy	FRP	Steel	FRP	FRP	CFRP
Frl	0.86	0.92	0.29	2.0	1.88	1.33

Table 4.3 Example high speed patrol craft

Name	Banzan class (VITA)	Bahamas class	Svetlyak class	SAAR 4 class	Hayabusa class	Roussen class	Armidale class
Country	UK	USA	Russia	Israel	Japan	Greece	Australia
Builder	VT Shipbuilding	VT Halter Marine Inc	Almaz Shipbuilding Co	Israel shipyard	Mitsubishi Heavy Industries Ltd	Elefsis Shipbuilding and Industrial Enterprises SA	Austal
Year	1995	2000	2002	2003	2003	2004	2005–2007
Mission	Fast attack craft	Patrol boat	Patrol boat	Large patrol boat	Fast attack craft	Fast attack craft	Patrol boat
Complement	35	35	28	30	18	45	29
Displacement (t)	376	375	375	415	200	580	270
L (m)	56.3	60.6	49.5	58.0	50.1	61.9	56.8
B (m)	9	8.9	9.2	7.8	8.4	9.5	9.5
T (m)	2.5	2.6	2.2	2.4	4.2	2.6	2.7
Max speed (kt)	35	24	30	32	44	34	25
Main engines	4×MTU 20V538 TB93 diesel	3×Caterpillar 3516B diesel	2×M550 diesel	4×MTU12 V331TC92 diesel	3×LM500-G07 gas turbine	4×MTU MD16V538TB90 diesel	2×MTU 16V4000M70
Power (kW)	4×3,450	3×1,620	2×3,530	4×2,750	3×4,025	4×2,205	2×2,320
Propulsion	4×prop	3×prop	3×CP prop	2×prop	3 WJ	4 prop.	FP propeller
Structure	Al alloy	Al alloy	Al alloy	Al alloy	Al alloy	Al alloy	Al alloy
Frl	0.767	0.507	0.7	0.69	1.02	0.71	0.55

Moving Forward

Monohull craft design has moved forward rapidly since the middle of the twentieth century, leveraging the same innovations that have enabled the other HPMV types to come into existence—lightweight power plant, light alloy structural materials, computer systems for automated stabilization and improved efficiency of propulsion. Given a design brief, a naval architect can normally present an efficient craft for the client from the performance point of view. For commercial craft, the focus has moved to the design of internal outfit—the passenger spaces to attract custom, or the cargo space to make loading and unloading as efficient as possible. In addition, with a monohull the external form can be devised to be most impressive and pleasing to the eye, see Fig. 4.25 from Nice harbour and Fig. 4.26 showing the super yacht Pelorus. Other HPMV have a greater challenge in this respect.

Fig. 4.25 A collection of super yachts moored on the east side of Nice Harbour, January 2011

Fig. 4.26 Super yacht with fine lines—the "Pelorus"

The biggest forward challenge is one of environmental efficiency. By their nature, fast marine craft consume fossil fuel and emit CO_2, and so the development of power systems with lower emissions is important. The driver for this is most likely to come from further development of monohull fast ferries, with move towards alternative fuels; this can then be leveraged by the other applications such as super yachts based on the environmental sensitivity of the clientele who will be having the craft built.

Chapter 5
Hydrofoil Craft

Boats with Wings

The best way to increase boat speed is to lift it out of the water. Once this is achieved, water resistance will drop in proportion to the hull area lifted away from the water surface and wave disturbance will decrease, consequently improving sea-keeping quality.

A boat with submerged wings—called hydrofoils—mounted on struts below and to the side of the hull generates lift in the same way as an aircraft wing, and lifts the boat hull out of the water, as shown in Fig. 5.1. Since water density is far higher than air (about 800 times higher), hydrofoil dimensions can be much smaller than aircraft wings: small enough to be added to a monohull boat and do not incur difficult weight and drag penalties.

The idea to lift fast boat hulls out of the water originated in the nineteenth century with experiments by Thomas Moy. He towed a trial boat with three foils suspended below it along a canal near London in 1861 and achieved sufficient lift for the hull to rise out of the water [2-6]. That was not his main focus though he was using the boat as a means of observing the behavior of the foils, as it was easier to do this in water than in air. He found this a successful approach. We do not have a record of whether he successfully built a glider, as many Englishmen were experimenting with at that time, but at least he had some hydrodynamic success! Inventors patented many different ideas for blades and fins to control hull draft and to lift a boat hull from the water in the second half of the nineteenth century, but the ideas were mainly followed up at model scale rather than full-size boats. In 1894, the brothers M and L Meacham designed a boat with fully submerged foils and an incidence control feeler arm linked to the bow foils, but this did not go beyond the design stage.

In 1898, Enrico Forlanini started experiments with foils, aimed at improving the takeoff for aircraft over water. Forlanini was very successful with his craft that had a kind of ladder of foils on each side at both the bow and stern on a craft that was propelled by airscrews in 1906; see Fig. 5.2. This craft reached 38 knots on Lake

L. Yun and A. Bliault, *High Performance Marine Vessels*,
DOI 10.1007/978-1-4614-0869-7_5,

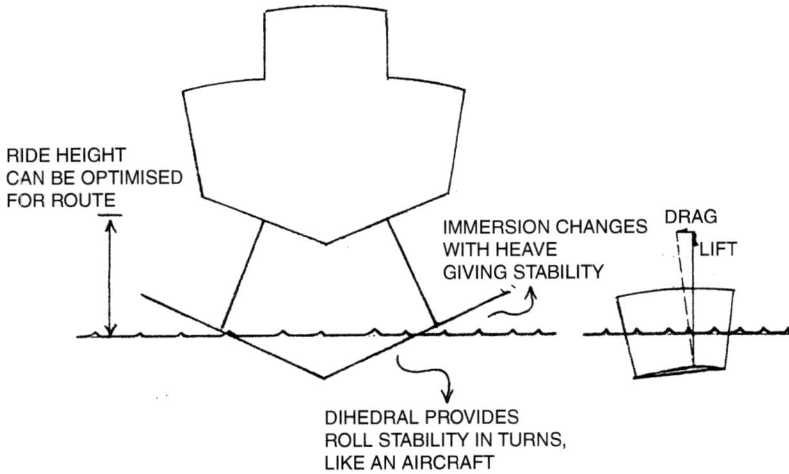

Fig. 5.1 Hydrofoil principle

Fig. 5.2 Forlanini hydrofoil from 1906 on lake Maggiore

Maggiore in Italy that year. Ladder foils proved to be a reliable form for early craft, though with the horizontal ladder steps the craft rise was not so smooth. This led to the idea to make the step foils in V formation so that the gradation of lift force was smooth as a craft accelerated and was lifted further.

This gave the idea for the simplification to larger foils that were canted. Also in Italy, A. Crocco and O. Ricanoldi built a boat with inclined hydrofoils, one V shaped at the bow and two more widely spread canted foils each side of the stern in 1907.

Fig. 5.3 Bell's Hydrodome HD-4 record-breaking hydrofoil

This craft was also driven by air propellers mounted on canted pylons containing drive shafts and right-angle gearboxes. The craft was 26 ft in length and achieved 42.5 mph power by its 100-bhp motor.

John Thornycroft in the UK, a builder of fast patrol boats, also experimented with foils under boats [5-1]. In 1909, he built the "Miranda III," a 22-ft-long boat intended as a planing craft, with a 60-hp engine. Planing was not initially achieved and so a foil was mounted under the bow. This served to lift the bow so that planing was achieved at around 15 knots, and the craft was able to accelerate up to 27 knots, a very high speed for the power installed.

Forlanini's experiments continued for a number of years, and in 1910 he had modified his hydrofoil boat to be driven by a water screw. Alexander Graham Bell visited Forlanini in that year and was taken for a trip on the lake, impressing him enough to make his own design of hydrofoil boat using Forlanini's system under license. Bell had also been experimenting with foils for seaplanes, and after his return to the USA worked on a series of designs culminating in a craft designated the HD-4, standing for Hydrodrome-4; Fig. 5.3. Running with two Renault engines, this craft made 54 mph. In 1919, he was able to install two aircraft engines of 260 kW each which enabled him to achieve a speed of 70.86 mph on 9 September 1919, a water speed record that stood for 10 years. Bell's associate Casey Baldwin later built two craft that he exported from the USA to the British Royal Navy in England for trials in the early 1920s. Unfortunately, they were tested in sea conditions along the southern coast of England that were too great for the design and so they did not survive the trials.

Fig. 5.4 Supramar PT-10 Freccia d'Orro

In 1927, Baron Hanns von Schertel began his own experimental work to improve the takeoff and landing performance of flying boats, an aircraft type that was gaining popularity for passenger flight at that time. Baron built seven different prototype boats to test foil configurations up until 1936. Initially, he concentrated on submerged foils, but finding it difficult to control steady immersion depth he moved on to surface-piercing geometries. The last test craft, with a V-form front foil and rectangular-shaped rear foils, was successful enough to perform a demonstration run on the Rhine between Mainz and Cologne. This brought an order from a shipping line for a passenger craft which was contracted for construction with the Gebruder-Sachsenburg Shipyard. It was decided to build one more prototype before the passenger craft to verify the design, and this craft was the 17-t VS-6 that achieved 47 knots. Before the commercial craft could be built, Germany was at war and so hydrofoils were designed and built for the German Navy for the next few years. Military hydrofoils as large as 80-t displacement and speeds up to 41 knots were designed and built. These were based on the surface-piercing Vee foil system of von Schertel.

After the war ended, von Schertel and Sachsenburg moved to Switzerland and founded Supramar A.G. to continue development of their surface-piercing hydrofoils. In 1953, the first commercial passenger hydrofoil, the PT-10 "Freccia d'Orro" (Fig. 5.4), was put into service on Lake Maggiore [5-2,5-3,5-4]. Later on, the craft was transferred to Lake Lucerne and operated there. In 1953 also, a larger PT-20 craft was built at the Lürsen Shipyard and named "Bremen Pioneer." This began a long line of successful designs, including much licensed production by the Rodriguez shipyard in Italy, Hitachi Zosen in Japan, Westermoen in Norway, and a limited production at Vosper Thornycroft in the UK.von Schertel improved the geometry of his surface-piercing foil system through adjustments to the plan form, and a patented system of air delivery down to the lower foils that steadied the lift force as the foils cut through waves. An example of the success of his designs is one of the craft built in WW2, an 80-t cargo craft with the ability to carry a 20-t tank with

Fig. 5.5 Alexeev Raketa hydrofoil

its supplies between Sicily and North Africa. The 105-ft craft powered by two 3,600-shp Mercedes Benz diesels was able to reach 37 knots in 1.5 m seas, an amazing achievement.

The Sachsenburg Shipyard was in the part of Germany occupied by Russia after the Second World War. The discovery of the advanced technology at the shipyard by Russian technologists encouraged the establishment of two development teams at home, one at the Admiralty Shipyard in Leningrad and another at the Sormovo Shipyard in Gorky on the Volga River. The logic in this Soviet period was for Leningrad to focus on designs for coastal operation in open water with surface-piercing or submerged foils, and the Gorky yard to work on craft for the Russian river system, a main transport artery past many towns and cities to both the Black Sea and via the river Don to the Caspian. It was in Gorky that Dr Rotislav Alexeyev and his team experimented with shallow-submerged foils in a number of configurations mounted under long, slender, monohull craft; Fig. 5.5. River conditions are relatively calm, so the "ride height" could be small; the declination angle of the propeller shaft for mid-mounted diesel engines could be low; and propulsion kept efficient. Speeds in the range 30–35 knots could be achieved with fully submerged foils. The first commercial craft was the 64-seat Raketa, at 28 m length, which began services from Gorky to Kagan in 1957 at speeds up to 32 knots. The craft was less expensive than an equivalent capacity ferry to operate while being speedy and giving a comfortable ride for passengers; so it became very popular, and led to a series of hydrofoils of increasing capacity being developed by the Sormovo team. Production of the Raketa alone has been as many as 400 vessels, built at a number of different shipyards in the Soviet Union. The work on shallow-submerged hydrofoils led to Alexeev thinking about its mirror image above the water in air in the 1960s, and developing the Wing In Ground effect craft we have described in Chap. 3.

Fig. 5.6 Russian river hydrofoil

In Leningrad, the focus was more on surface-piercing hydrofoils following the von Schertel concept, though adjusted for service in lower sea conditions in the Baltic, and the lower Volga river, with lower ride height. It is clear from the sheer number of craft built by these two shipyards over the second half of the twentieth century—several hundred—that they are efficient for their mission, and have satisfied the passenger clientele. A number of the later models have also been exported to Greece for use on the Aegean, and to Western Europe for use on the Danube and Rhine. While the original diesel engines have proved less reliable than the European equivalent, replacing the engines has often solved that problem. The structure on these craft is resilient and the accommodations spacious, so ferry services outside Russia have also proved a success over many years. In fact, there are many in service in Russia and Europe after more than 25 years' service after having been re-engined, and reoutfitted, examples being fleets in Budapest, Vienna, etc. (see Fig. 5.6).

At the end of the chapter, there is a table with basic details of the craft series built to Supramar design, as well as the various models built in the USSR. While both of these concepts and vessel series were well suited to their respective markets, in the rougher seas of the Atlantic and Pacific coastlines the challenge was more difficult.

The potential of the deep-submerged foil was realized initially through development of military craft for the Italian, US, and Canadian Navies, and later specifically by Boeing for passenger service with the Jetfoil design that was later licensed for production in Japan to Kawasaki. First, we discuss in a bit more depth the hydrodynamics of foil systems, and then describe some more of the main craft developments to date.

Before we go further, a quick recap on the key design characteristics to be built into a hydrofoil.

- The foil system needs to provide the forces to lift the hull out of the water to sufficient height to clear the tops of waves in the intended service sea state.
- The foil lift has to balance the craft weight and also to balance through the craft's center of gravity, and provide righting forces with roll or pitch.
- The balanced ride height has to be statically stable, either through automatic change of lift force due to changing immersion or by automated operation of control surfaces.

There are a number of different ways to achieve this, depending on the sea state the vessel has to operate in. We start with river and lake operations (relatively calm waters), coastal operations, and lastly open sea conditions.

Shallow-Submerged Hydrofoils

Hydrofoils have certain basic characteristics that are important to the designer. The lift reduces as the foil approaches the water surface while below an immersion greater than the foil chord the lift is constant. Also as a foil approaches the water surface, there is a tendency for cavitation on the upper (low pressure) foil surface which can have damaging effects similar to the effect on marine propellers. Hydrofoils with surface-piercing lifting foil arrays minimize this by a combination of design accounting for the cavitation for the surface-piercing section, and employing fence structures to protect the lower foils.

A hydrofoil cross section has a sharp leading edge to encourage cavitation over the upper low-pressure surface and create a clean cavity to avoid pitting damage, and relatively shallow camber. Foil sections can be designed with higher camber as water depth increases and with thicker section to take the hydrodynamic load. In contrast to an aircraft wing with reducing chord out to the wing tips, a hydrofoil may actually have wider chord towards the tip so that lift forces are higher and assist transverse stability in heave and roll.

The challenges of hydrofoil dynamic design are similar to the aerodynamic challenges for a WIG craft (see Chap. 3) but in reverse, rather than mirrored about the surface. The lift degradation as foils approach the surface gives the hydrofoil craft quasi-static stability or inherent stability in heave and pitch; so in case the craft moves down from the balanced water line, the craft will be lifted automatically due to increase of lift, and vice versa. Surface-piercing foils accentuate this property as the part of foils normally in air also give additional lift as a craft heaves or pitches downwards. Such craft can be operated without an automated control system using flaps or elevators.

Using the shallow submergence effect, hydrofoils with shallow dihedral can give inherent stability so that the craft can run at a stable elevation with the hull above

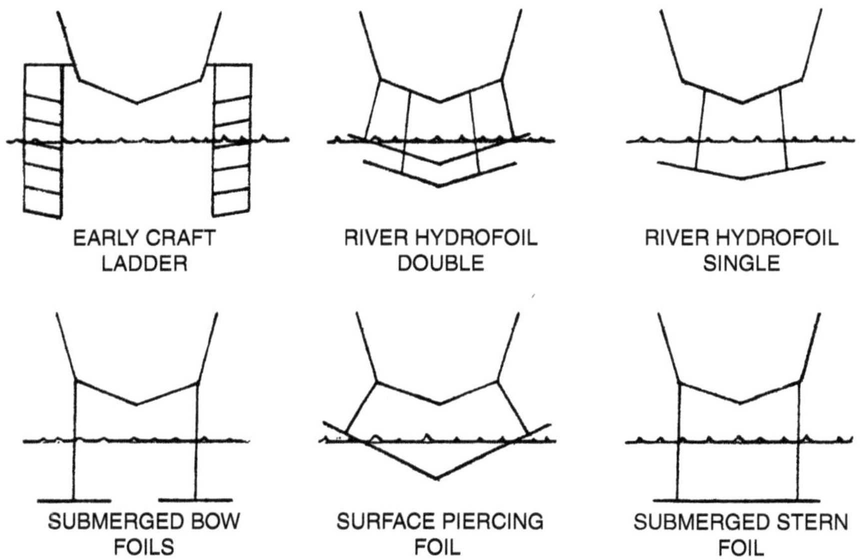

Fig. 5.7 Hydrofoil configurations

the water. In fact, this not only gives the craft positive vertical stability, but can also help it with both transverse and longitudinal stability. Figure 5.7 shows several configurations of hydrofoil craft with inherent stability. Figure 5.7b shows a surface-piercing hydrofoil with positive transverse stability. Since the transverse cross section of the hydrofoil is a V profile, when the craft rolls to the right, as shown in Fig. 5.8a, the area of foil immersed on the right side will be increased generating a restoring moment to keep the craft stable.

The shallow submergence effect is similar to the surface effect for aircraft wings close to the ground, especially so as the effect is reversed (lift decreases as the foil approaches the surface) and is effective only for immersion between zero and half the chord of the hydrofoil section. Thus, if the foil is 0.5 m width, a typical dimension, then the surface stabilization is effective for immersion in the range 0–25 cm. This is not a problem for river navigation as needed for the craft in Russia along the Volga and Don system, but not a help for offshore navigation.

In the case of a submerged foil with an aerofoil wing configuration, it also has a positive inherent stability due to increasing submerged depth at the right side so as to increase the lift at right side and generate a restoring moment, as shown in Fig. 5.8b. This moment is not as strong as for the surface-piercing foil, and so active elevators to further increase the turning moment are important for fully submerged hydrofoil craft.

The river hydrofoils developed in Gorky have foils at bow, stern, and also amidships. Apart from the hydrodynamic stability, this arrangement assisted the structural design of these rather long hulls. Later models, such as the Meteor and Kometa,

a

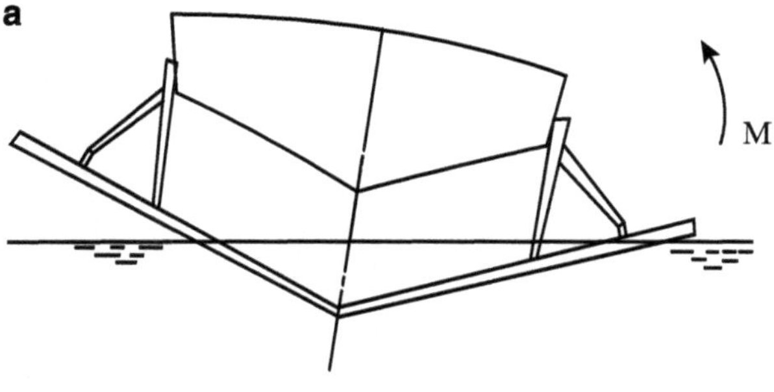

b

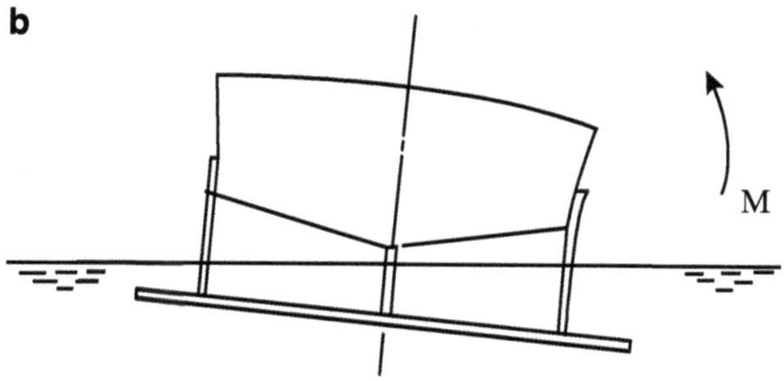

Fig. 5.8 Hydrofoils rolling in turn. (**a**) surface piercing. (**b**) submerged foils

also had biplane foil arrangements, the upper foils riding out of the water when at cruising speed.

With respect to maintaining a positive longitudinal stability, there are three types of foil arrangements, as shown in Fig. 5.9, where (a) shows the aircraft layout with forward foil as the main lifting foil supporting most of craft weight, and a tail foil supporting a smaller part of the craft weight and acting as stabilizer for longitudinal motion similar to the tail plane in an aircraft. The craft's longitudinal center of gravity should be near but a little aft of the main foil. Figure 5.9b shows a tandem arrangement of foils, with almost equal load for both forward and rear foils. The craft CG for this arrangement is centrally located between both foils in longitudinal direction. Finally, Fig. 5.9c shows the canard-type arrangement of foils, i.e., most part of load concentrated at the rear foil, and only a little part of craft weight supported by forward foil, so the rear foil is the main foil and forward foil is an auxiliary foil primarily functioning for stability. In this case, the craft CG is located just forward of the rear foil between the two foils. This arrangement improves seakeeping compared to the aircraft or tandem layout, since the influence of waves is higher on a bow foil than a stern foil.

Still another configuration is that where the craft has only one foil, i.e., a main foil arranged forward for supporting most of the craft weight, but a small part of the weight supported by the hull planing at the stern, as shown in Fig. 5.10. The small bow foil at bow shown in the figure is helpful during takeoff from displacement to planing operation. During takeoff, the added lift acting at the bow foil causes the bow part of craft to clear the water surface quickly and also increases the trim

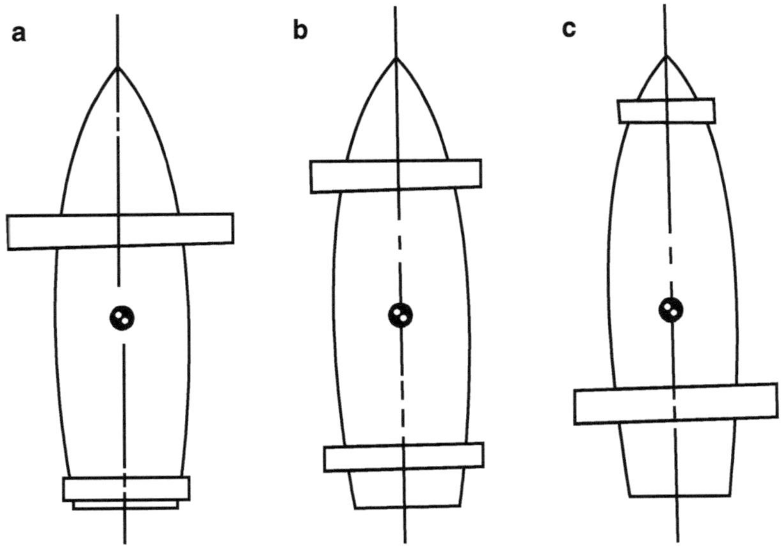

Fig. 5.9 Shallow-submerged foil arrangements for longitudinal stability

Fig. 5.10 Single main lifting foil configuration—Bras D'Or

angle so as to increase the hull planing lift and accelerate the takeoff operation. In addition, this small foil can improve seakeeping, particularly in head seas.

There is an interesting story relating to development of this type of craft. A prototype hydrofoil craft with tandem foil arrangement was being tested in the USSR during the 1950s and experienced a broaching motion in the following waves.

Fig. 5.11 Chinese Coastal Attack Craft with single bow foil

The craft suddenly broached with the stern rotating forward around the bow until the craft changed direction by 90°, ending beam on to the seas. It was a rather dangerous phenomenon, and also unacceptable to the crew. When the incident happened, the rear foil was damaged, and in order to quickly repair the craft for testing it was decided to remove the rear foil so that trials might continue. Fortunately, the broaching phenomenon did not recur, so the engineers decided to leave out the rear foil in later batches of these hydrofoil craft. The subsequent testing was successful, verifying the foil performance and showing that the vessel performance was acceptable for the ferry service that it was designed. For operation along the great rivers of Russia and USSR where the surface was nearly calm most of the time, this was a practical choice!

Figure 5.11 shows a Chinese hydrofoil torpedo boat with a single main foil, and an auxiliary bow foil for improving seakeeping at high speed, which was designed and operated in the 1960s. It is interesting to consider this craft against the 'Miranda' built by Thornycroft in 1909 that we described above on page 163!

Figure 5.12 shows a Russian surface-piercing shallow-submerged hydrofoil "Voskhod," with displacement 28.4 t (full loaded), 20.4 t (light), speed, 60 km/h, range 500 km, 71 passengers, power plant one M401A-1, 735 kW diesel, driving a water propeller. Hull and superstructure are made of aluminum/magnesium alloy. The hull is an all-welded structure, and the riveted superstructure is made by resistance spot welding on glue due to the thin plate.

Figure 5.13 also shows a sea-going passenger, self-stabilized, shallow-submerged, surface-piercing hydrofoil of design series "Kolkhida," running along the coastline, with displacement 74 t (full loaded), 56.9 (light), operation speed 35 knots, range 200 nautical miles, 155 passengers, 5 crew, 2 MTU diesel with total power of 2,520 kW, driving two water propellers. Its foil-borne operation can be in up to 2.5 m wave height (Russian standard at the average of 3% max wave height). The hull of Kolkhida is made of aluminum alloy. Since it was to be operated in open seas, in order to ensure its seakeeping quality in rough sea, an automatic stability

Fig. 5.12 Voskhod

Fig. 5.13 Kolkhida

augmentation system is installed in this design. This is an automatic system for controlling the main foil flap angle and feeding air ventilation on the back of the foil to regulate the main foil lift in waves.

Figure 5.14 shows a similar craft, but larger, the "Olympia-M," from Russia, with full displacement 138 t, speed 38 knots, 250 passengers, 2 diesels with 2,000 kW each, driving two water propellers, which can also operate in waves of 2.5 wave height (average of 3% max wave height). The foils were arranged in aircraft configuration, with large surface-piercing forward main foil supporting most of the craft weight and a small flat stern foil for pitch stability.

Fig. 5.14 Olympia-M hydrofoil

Surface-Piercing Hydrofoil Development

von Schertel continued the work begun in Germany once established with Supramar in Switzerland in 1952. Introduced above, first came the 32-passenger PT-10 in 1953, with its V-shaped bow foil, and flat rear foil that also supported the outer end of the transmission shaft to the propeller from the centrally mounted diesel motor. This arrangement gave a characteristic layout to the subsequent PT series, with passenger salons forward and aft, a central engine room, and the wheelhouse mounted above this. A number of PT-10 craft were built, mainly for cross-lake commuting and tourist sightseeing trips. The PT-10 was suited to service on lakes or rivers but not to open coastal conditions. This led to the PT-20, a larger 72-passenger craft with a bow foil unit that could be rotated by a hydraulic ram so as to adjust its incidence to accommodate to varying passenger load distribution and the sea conditions.

In Italy, there was an opportunity available for a fast ferry service between the mainland and Sicily across the Messina strait. This attracted the Leonardo Rodriquez Shipyard, now known as Rodriquez Cantieri Navali SpA, who took out a license with Supramar and built the first PT-20 in 1956 (Fig. 5.15). In service, the hydrofoil ferry was found to operate at breakeven, charging a reasonable surcharge above the traditional ferries with as little as ten passengers aboard between Reggio on the Mainland and Messina in Sicily. The quicker journey made the craft so popular that it ran almost full for all its trips. This success led to many orders for PT-20 from other ferry companies around the world. By 1958, Rodriguez had built 20 craft, and by 1973 over 70 of the type.

In 1958, Supramar began its next step upwards in scale with the PT-50 (Fig. 5.16), a craft with accommodation for 105 passengers in the first version and between 135 and 160 in the Mk II version. PT-50 was powered by two of the Daimler Benz 12 cylinders, 1,100-shp diesels that powered the PT-20, each driving a subcavitating propeller through its long-angled shaft. The foil arrangement was altered a little,

Fig. 5.15 Supramar PT 20

Fig. 5.16 Supramar PT 50

back to a fixed installation as on PT-10, but with flaps fitted to the bow foils to allow for trimming. Load sharing was 58% forward and 42% aft.

The PT-50 was certified for coastal operation by many of the world's marine classification societies. It has been put into service on routes across the British Channel, the Mediterranean, in Japan, and South America and has become probably the most widely used hydrofoil ferry to date. Such craft as PT-20 and PT-50 have seen heavy use in service. Apart for the normal maintenance of the highly utilized diesel engine, the main issues for a hydrofoil operator are marine fouling and physical damage to the foils and hull surface, and cavitation damage to the propellers. While these are also important for any fast marine craft, the performance degradation for a hydrofoil is more extreme; so operators have to take particular care to keep their craft "clean." Since this means getting at the submerged parts of the craft, a suitable dock or crane lift is important within practical reach of the service itself.

Fig. 5.17 Rodriquez RHS 160

Rodriguez continued expanding the range with its own hydrofoil designs assisted by Supramar, the RHS 70 (71 passenger) from 1972, RHS 110 (97 passenger) from 1971, RHS 140 (150 passenger) from 1971, RHS 150 (150 passenger) from 1980, RHS 160 (140–200 passenger) from 1974, and RHS 200 (250–400 passenger) from 1981 (Fig. 5.17). All of these craft series were based on the Supramar surface-piercing foil system. Leading particulars can be found in Table 5.1. Supramar supported Rodriguez and also its other licensees Westermoen, Vosper, and Hitachi to design and build hydrofoils to their design [5-5]. In parallel with Rodriguez in the 1960s, Supramar developed the PTS-75 that was built by Vosper Thornycroft, and PT-50, 75, and 150 built by Westermoen. It was Westermoen that began the introduction of stabilization by air entrainment on the foils.

Following some trials with a small test craft in 1967, Westermoen installed a separate air bleed stabilization on each of the port and starboard sides of a fully submerged aft foil fitted to one of its PT-50 craft. The system works with a flap that allows air to be sucked onto the upper surface of the foil through a duct in the support strut to vary the lift force. The amount of air is controlled by a damped pendulum and a rate gyro installed in the hull. More air is admitted to the foil that is less submerged (up) and less to the more submerged (down) so as to provide a righting moment in roll. As the foil passes through waves, this smoothens out the lift variation and gives a steadier ride. This is very effective in both head and cross seas.

Trials with the PT-50 showed a reduction of up to 75% in motion response in roll, a really big improvement, once the air volume admittance was optimized. Initial setup of the system did give a speed reduction in calm conditions, but later development minimized this while the performance in a sea state was improved considerably, allowing operators more confident timetable scheduling. Westermoen continued to refine the ride control system design while Supramar developed the PT-150, a craft for 250 passengers. Westermoen built two of these craft for service

Table 5.1 Example commercial hydrofoils

Name	Tayfun	RHS 150	RHS 200	Foilmaster	North Star	Mountain Ge-lu
Country	Russia	Italy	Italy	Italy	China	China
Delivery year (first)	1969	1980	1981	1994	1994	1996
Builder	Almaz shipyard	Rodriquez	Rodriquez	Rodriquez	n/a	n/a
Passengers	250	150	210–238	242	294	82
Displacement, t	135	65.5	130	112	118	32
Loa, m	42.55	28.7	35.8	31.2	29.1	22.96
Hull beam	8.3	5.6	7.6	6.78	8.6	4.8
Boa, m	n/a	11	14.5	13.3	9.2	7.6
Draught, m:						
Foil borne	4.6	3.1	4.55	3.89	4.5	2.05
Hull borne	2.0	1.4	2.05	1.45	n/a	1.1
Machinery	2×MTU 16V396TB84	2×MTU 12V331TC82	2×MTU 16V652TB71	2×MTU 16V396TE74	Allison-501KF, gas turbine	KTA19-M2
Power, kW	2×1,900	2×1,052	2×1,914	2×1,550	2×2865 (max 3,180)	2×461
Speed, knots:						
Full	40	35	37	37	50	38
Operating	37	32.5	35	35	43	37
Range, nm	295	130	200	150	100	216
Seakeeping	SS4	SS4	SS4	SS4	SS4	SS4
Hull type	Monohull	Monohull	Monohull	Monohull	Monohull	Monohull
Structure	Al Alloy	Al Alloy	Al Alloy	Al Alloy	Al Alloy	Al Alloy
Propulsion	Propeller	Propeller FPP	Propeller FPP	Water jet	Water jet	Propeller FPP
Foil type	Surface piercing Plus auto Automatic control	Surface piercing	Surface piercing	Full submerged Plus automatic control	Full submerged Plus auto Automatic control	Surface piercing
Foil load Fore/rear	n/a	60/40	60/40	n/a	30/70	52/49
Frl	1.01	1.07	1.02	1.09	1.52	1.3

Name	Westfoil	MEC-1	Jetfoil 929	Foilcat	Rainbow SS-400	Hayate TSL-F	FSH 38
Country	USA	Italy	USA	Norway	Japan	Japan	Italy
Builder	Westport shipyard	Rodriquez	Boeing	Fjellstrand	Mitsubishi, Shimonoseki	Kawasaki, Kobe	Rodriquez
Delivery year (first)	1991	1992	1980–1990	1995	1993	1994	2010 2 prototypes
Passengers + crew	149	146	450 + 8	403	341 + 4	14 crew	245
Displacement, t	89.4	n/a	110	n/a	75 (35 dwt)	n/a	140
Loa, m	25.6	25.0	27.4	35	33.24	17.1	37.25
Hull beam	7.6	6.7	9.5	12	12.0	6.2	8.0
Boa, m	8.0	8.4	11	12	13.2	6.36	8.0
Draught, m:							
Hull borne	4.75	2.8	5.0	4.7	4.5	3.1	4.3
Foil borne	1.83	0.95	2.0	2.55	2.1	1.6	
Machinery	4 DDC 12V-92TA	2 MWM TBD604 BV8	2 × Allison 501K20A GT	2 GE-LM500 GT	4 MHIS 16R-MTK	Single gas turbine	2 MTU 16V4000M70
Power, kW	4 × 625	2 × 840	2 × 2,450	2 × 4,026	4 × 2,100	1 × 2,820	2 × 2,320
Speed, knots:							
Maximum	50	38	59	52	45	41	47
Operating	42	34	50	45	40	41	45
Range, nm	730	200	405	300	n/a	n/a	430
Seakeeping	SS 4	SS4	SS5 (3.65 m max)	SS4	SS4	SS5	SS6
Hull type	Monohull	Monohull	Monohull	Catamaran	Catamaran	Monohull	Monohull
Structure	Al Alloy	Al Alloy	Al Alloy	Al Alloy	Al Alloy	Al Alloy	Al Alloy
Propulsion	2 × Ducted air propellers	Hydraulic-driven Rexroth propeller pod	2 × Waterjets Rocketdyne 20	2 × Waterjet	2 × Waterjet	2 × Waterjet Hamilton	2 × FPP or 1 × contrarotating podded unit
Foil type	Fully submerged Automatic control	Surface piercing Plus control	Fully submerged Plus auto Automatic control	Fully submerged Plus auto Automatic control	Fully submerged Plus auto Automatic control	Fully submerged Plus submerged buoyancy	Fully submerged Plus auto control
Frl	1.63	1.24	1.86	1.43	1.28	1.63	1.26

Fig. 5.18 RHS 200

between Copenhagen and Malmö across the Kategatt. On the PT-150 (Fig. 5.18), air-entrained stabilization was installed on the center part of the front foil as well as the stern foils. This system improved pitch and heave motion as well as the roll damping from the rear foil system. The bow foil air admittance system is controlled by an accelerometer and rate sensor. The bow foil also has flaps on the outer part of the foils for craft trim and assistance during takeoff.

Westermoen's hydrofoil developments were driven by the more difficult sea conditions on the Norwegian coastal routes [5-6]. While successful in Mediterranean conditions, these more challenging routes meant that the ride control developments were very important to provide a satisfactory service performance of the craft, both for passengers and the ferry companies. Characteristic of many technology developments, the communication between operator, shipyard, and the designers was often a challenge due to the different expectations. Westermoen's achievements were very significant, perhaps more so than the designers at Supramar appreciated. Unfortunately, for the shipyard, the operators had their own challenges to keep regular service; and so although for a while in the 1960s the hydrofoils created a new market for rapid passenger transit, once the catamaran became a competitive alternative for these routes in the early 1970s, both the operators and Westermoen moved over into this new market.

In Japan, Hitachi built PT-20 and PT-50 craft in series from 1961 to the early 1980s, a total of 8 PT-20 and 15 PT-50, the latter craft with dynamically controlled flaps in the front foils to damp roll and pitch motion. All these craft entered service on coastal ferry routes in Japan and performed reliable service for many years. The surface-piercing hydrofoil has advantages and limitations, however, as we now summarize.

Advantages and Disadvantages of Surface-Piercing Hydrofoils

Advantages

- *High-speed* craft with surface-piercing foils can be designed for speeds up to 40 knots, and give low fuel consumption compared to monohull craft. Above this speed, cavitation of both foils and propellers reduces the efficiency and limits economy; see below.
- *Low wave making* at high speed, so causing low disturbance to boats nearby the craft track and minimizing waves' breaking on river banks, as shown in Figs. 5.5 and 5.6; so the design is useful for inland operations.
- *High maneuverability in restricted waterways* due to small overall size, compared with CAT, WIG, ACV, and SES for same payload. The hydrofoil configuration can be added directly to a simple planing monohull.

Disadvantages

- *Foils protrude outside the craft beam*, and make docking the craft alongside a pier a more complicated operation. An operator has to invest in special equipment for fendering and access ways to the boat for passengers, increasing the terminal investment.
- *Cavitation* occurs when the foil is moving at high speed, and this limits economic service speed for such craft to below 45 knots. The "cavitation" is the main obstacle resisting the development of this type of craft. In general, even submerged hydrofoils are practically limited to about 60 knots, as we explain below.
- *Size limitations*: Since foil lift increases with size to the second power and craft weight increases with size to the cubic power, in order to be balanced, the craft speed should be increased on the square root of the linear size. Thus, when scaling up craft size, the speed should increase on the square root of the scale; however, there is limited speed increase possible due to the "cavitation" obstacle. In short, the surface-piercing hydrofoil craft is difficult to extend over 200–300-t displacement due to the practical speed limitation of 40 knots.
- *Seakeeping*: Since foils of this type are located close to the water surface, they are disturbed by waves, leading to heave and pitch motions and discomfort for the passengers and crews in high sea states. To ensure safety, including avoiding possibility of broaching in following seas, the operating envelope has to be strictly controlled for offshore operations.
- *Propulsion system limitations*: Water jet propulsion with high propulsive efficiency at speed is difficult to incorporate into a hydrofoil due to the hull being clear from the water surface compared with monohull craft, CAT, SES, etc. This problem has been partly solved for the fully submerged foil craft but has not proven attractive for the application to surface-piercing hydrofoils possibly because of the higher vertical movement of the rear foil in a seaway and problems this brings to design a reliable water-jet intake.

How can the disadvantages above be minimized? The PT series of craft were all designed within an envelope controlled by the power system: high-speed diesels from Daimler Benz, or MTU, simple vee drive, and long shaft transmission to subcavitating propellers. This meant service speed in the range 30–40 knots depending on payload. While it would be possible to install gas turbines with much higher power rating and lower weight, the use of this power would require a large water jet or supercavitating propeller system. Both of these technologies would take normal ferry operators outside their "comfort zone," and place considerable technical and economic demands on the ferry operation. Hoverlloyd set up such a more radical operation for hovercraft ferries in the 1970s from scratch at Pegwell Bay in the UK, but this is a very special case. Ferry operators are generally very conservative, and do not adopt new technology unless it is simple and reliable, or the economics are a no brainer, as was the case with the PT-20 and PT-50.

Both Supramar and Rodriguez worked on gas turbine-powered designs in the 1970s and 1980s. Rodriguez did design and build the Maximum Efficiency Craft (MEC)-1. This had the two diesel engines mounted right aft, with Rexroth hydrostatic drive through vertical fins to two propellers mounted in front of the rear foils that take the majority of the weight, reverse of the normal Supramar setup. The MEC-1 for 146 passengers was aimed at a 38-knots service speed, and carrying capacity increased by 25% compared to the equivalent standard design. While Rodriguez continued to have such designs available, the ferry market began to change in the early 1990s, as large car-carrying catamarans were introduced. Fleet replacement, therefore, turned to other craft types rather than to next-generation surface-piercing hydrofoils.

With the PT-150, Supramar had developed its craft up to essentially the maximum that cross channel (short distance) passenger-only ferry services demanded, and with service wave height up to 3 m significant this was more than enough for most applications. The next technology step was already being pursued in the naval world during the 1960s and was able to move across to passenger ferries in the 1970s—the fully submerged deep draft hydrofoil.

Before we move on to this craft type, mention should be made of the hydrofoil patrol craft developed in Canada, the FHE-400 "Bras D'Or" (Fig. 5.19). A prototype constructed by De Havilland Aircraft Company, this craft was intended as the start of a fast coastal patrol class. Powered by a Pratt and Whitney gas turbine of 22,500 shp, the hull with the lines of a frigate, this 260-t craft was able to run at 50 knots in 3.5 m seas or close to 60 knots in calm water. Its main foil just aft of amidships carried over 70% of the weight. It was a complex design with a lower fully submerged section attached to the vertical struts carrying the Z drive from the gas turbine to supercavitating propellers aft of the foils. Upper foils at 45° angle came down from the hull to meet the lower foils. Both had flaps fitted. At the bow was a steerable foil in diamond formation. The craft certainly had high performance as an interceptor craft, and was at the edge of technology in the early 1970s. Unfortunately, the oil shocks of the 1970s meant that this craft did not go further, rather like the SES3000 program. The use of gas turbines and supercavitating propellers proved what was possible with surface-piercing foils, but could there be a system that would be more efficient?

Fig. 5.19 FHE-400 Bras D'Or

Deep-Submerged Hydrofoil Craft

We have listed the advantages and disadvantages of surface-piercing hydrofoils. The stabilization system of Supramar gave a satisfactory ride for coastal ferries, but for military craft the desire back in the 1960s was for an ocean-going patrol craft that could outrun a high-speed submarine, part of the intent behind the FHE-400. Large patrol submarines in both Soviet and NATO Navies were able to reach speeds in excess of 40 knots, and were unaffected by sea state. The possible answer was the fully submerged hydrofoil, with a stabilization system relying on Doppler sonar sensors of the wave surface ahead of the craft to operate control surfaces at the foils in a similar way to an aircraft.

In the 1960s, research began in earnest in the USA and in Europe. The USA sponsored competitive programs at Grumman and Boeing while the Canadian Navy worked with DeHavilland on the FHE-400. Grumman's work began in 1957 with a US Maritime Administration (MARAD) research contract to look at open ocean hydrofoils for fast freight and passenger service in the size range 100–3,000 t, and speeds above 50 knots. Its initial studies resulted in a design for a 95-t trial craft, the 105-ft "Denison" which actually used surface-piercing forward foils carrying 85% of its weight and a fully submerged T foil at the stern (Fig. 5.20). The craft was powered by a General Electric J-79 jet engine marinized by addition of a separate free power turbine by GE giving 14,000 shp to drive a supercavitating propeller. The turbine was mounted in the stern with a Z drive arrangement down through the foil

Fig. 5.20 Grumman Denison

strut to the propeller, a major technical achievement at the time. It had a stabilization system based on aircraft technology with gyro and rate sensors controlling flaps at the foils provided by Hamilton Standard, and achieved speeds up to 71 mph in calm conditions and service at reduced speed around 55 mph in 2.6 m seas. The Denison trials were very successful, but at the end, the US Navy who shared the program with MARAD decided to go a different direction and removed support; and MARAD decided to terminate the program and not to pursue development of commercial hydrofoils on its own.

Grumman was able to continue with its hydrofoil development on two parallel tracks. On the commercial side, it developed a design called the Dolphin, a 64-t hydrofoil ferry for 116 passengers with fully submerged hydrofoils powered by a 3,600-shp Rolls Royce Tyne gas turbine driving a KaMeWa controllable pitch supercavitating propeller through a Z drive arrangement (Fig. 5.21). The rear T foil and forward pair of T foils were all designed to fold upwards for maintenance or shallow water boating. When boating, two GM diesels each provided 216 shp through water jets for propulsion. The Dolphin had a service speed of 48 knots and could continue service in up to 3 m seas. Dynamic control of craft level and motion was by an autopilot provided by Garrett comprising a forward-looking height sensor, gyros, and accelerometers linked to a computer system which controlled the incidence of the foils. The craft was built by Blohm and Voss in Germany and first put in service in the Canary islands, but had reliability problems and after less than a year was returned to the builders. It was next tried between Miami and Freeport, Bahamas. The sea conditions again made services difficult and so this operation came to a halt. In 1969, it was moved to the Virgin islands for a summer season, and finally sold to the US Navy. Clearly, just then, the design was ahead of its time and the equipment was pushed to its limit.

The main challenge in this period, common with the development of fast monohulls, was propeller and foil performance in conditions of cavitation. Supercavitating propellers aim to have a steady environment by creating a cavity

Fig. 5.21 Grumman Dolphin

over the whole top (low pressure) blade surface. The problem is that when close to the surface and in variable conditions such as high seas, the cavitation may not stay steady, so causing vibration and localized erosion on the blades. This can also be a problem for the lifting foils. A solution for monohulls, particularly race boats, has been the ventilated or "surface riding" propeller. The problem for early submerged foil hydrofoils was unsteady performance in a seaway, an uncomfortable ride for passengers, and high maintenance costs. The solution was found by intensive research to perfect submerged foil geometry, mainly under the Navy Programs that were carried out in the 1960s and 1970s. The propulsion issue was solved in the USA through adoption of the water jet for propulsion.

Naval Hydrofoils in the USA

Grumman's second track was a design brief for the US Navy for AGEH-1, an Auxiliary General Experimental Hydrofoil, which was awarded in October 1961 [5-7]. Grumman teamed with Newport News Shipbuilding and GE to develop the 320-t, 64-m-long craft aimed at higher than 50 knots. After initial design, the $17-million construction proposal by the group was $5 million higher than the USN target, so they rebid for construction. Lockheed Shipbuilding in Puget Sound won the contract in July 1963 with a different supporting subcontractor team. The keel was laid in May 1964, and the ship launched in June 1965. Foil-borne flight was not until March 1968 after outfitting, and commissioning by the Navy was not complete until March 1970. While strikes at the shipyard had a lot to do with the delays, the

Fig. 5.22 AGEH-1 plainview

ship also had considerable technical problems. One can wonder if the transfer from the original development team to a completely different contractor had an influence on this outcome.

The Plainview (Fig. 5.22) did achieve speeds over 50 knots powered by its two GE LM-1500 15,000-shp marine gas turbines, each driving two titanium supercavitating propellers, including running in 3 m seas. The propulsion transmission from the gas turbines was by a Z drive through the main foil struts. These were mounted just forward of amidships and took 90% of the craft weight, very much an aircraft configuration. The submerged foils were designed by Grumman to take subcavitating foils to their limit, so had significant sweep.

Like the Boeing-built High Point (see below), Plainview had to have a program of corrections before it began more reliable operation for the US Navy, but eventually it was able to participate in trials, including torpedo and missile launch, remotely piloted vehicle (RPV) operations, and much more. The craft was decommissioned in September 1978. While 8 years may not seem a long time, as a trials craft this was certainly a meritorious service.

Following the design of Plainview, Grumman moved on and designed the PGH-1 Flagstaff fast gunboat for the US Navy (Fig. 5.23) based on the experience from the Dolphin, as part of a competitive development program awarded in April 1966. Flagstaff was a 67.5-t displacement craft of 22 m length powered by a Rolls Royce Tyne Gas turbine of 3,550 shp, giving it a maximum speed of close to 50 knots. The second prototype PGH-2 was designed and built by Boeing. PGH-1 was very similar to the Dolphin in specification, though it would appear that the "bugs" that were such a difficulty for Dolphin were ironed out for Flagstaff. The hydrofoil was

Fig. 5.23 PGH-1 Flagstaff

commissioned in November 1968, and after a year with the Navy Pacific Fleet she was delivered to Vietnam for river patrol during "Operation Market Time." In 1970, she returned to San Diego and continued Navy service until being loaned to the US Coastguard for trials as a Cutter for 3 months. In September 1976, the USCG again took the Flagstaff for a longer evaluation in the East Coast 200 mile economic zone; this lasted until September 1978. The evaluation established the foundation of experience for the US Navy Patrol Hydrofoil class later procured.

Boeing Marine Systems started its hydrofoil developments in June 1960 when it won a contract to build a 120-t hydrofoil to test antisubmarine warfare from the US Navy. The craft was aimed at greater than 40 knots' service speed, and was fitted with tracking sonar and armed with homing torpedoes. While the Grumman craft had two foils taking most of the weight forward in aircraft formation, the PCH-1 "High Point" (Fig. 5.24) had a single forward T foil and twin aft T foils in the stern quarter either side of the hull supporting 70% of the craft weight, the canard formation. Two Rolls Royce Proteus marine gas turbines of 3,900 shp each drove contra-rotating subcavitating propellers through a Z-drive transmission to the nacelle of the rear foils. The concept design for PCH-1 had been developed within the US Navy's design bureau early in 1958. Plans and detail specifications were developed between mid 1958 and January 1960, following which construction was tendered. Boeing won the order in June 1960 and constructed the craft through a subcontract with Martinac Shipbuilding. The keel was laid in February 1961, with launch in August 1962, handover for commissioning in November 1962, and Navy receipt in August 1963.

Fig. 5.24 Boeing PCH-1 high point

The US Navy's initial experience with High Point was somewhat of a trial. Protective coatings applied to the foils came away and caused hydrodynamic problems while the design of both the propellers and foils proved to have cavitation problems, and so needed adjustment. The height sensor also operated erratically in high winds and required redesign to stop the craft porpoising in certain sea conditions. Boeing itself invested in small research craft to provide test data for foil design, controls design, and later also design of the water jet system that was installed in the PHM and Jetfoil craft. The US Navy also sponsored Boeing's construction of the FRESH-1, a catamaran test vehicle propelled by a jet engine that was able to investigate hydrofoil performance at up to 115 mph (100 knots).

Testing with High Point continued in the US Navy until 1965 when it was passed over to the Hydrofoil Development office of the David Taylor Model Basin. High Point was used as a test platform investigating foil hydrodynamics, motion response, and performance correlation over the period to December 1984 when the vessel was deactivated.

Boeing's experience with its water-jet test craft, "Little Squirt," gave it the confidence to employ water jets for its competitor in the Patrol Hydrofoil field, PGH-2 Tucumcari (Fig. 5.25). This craft of 70 ft length and 57 t displacement was ordered in 1966 and delivered in March 1968 to the Navy in San Diego, being joined by PGH-1 from Grumman later that year. Both craft were sent to Vietnam for service and comparative evaluation as coastal patrol craft. On return, Tucumcari was delivered to Europe for a NATO participation tour from April to October 1971. Following this, she was deployed to the Atlantic fleet, but a year later was badly damaged on a coral reef offshore Puerto Rico and was not able to be returned to service. By this time, Boeing had sufficient data to support future craft design being centered round water-jet propulsion and a canard arrangement of foils.

Fig. 5.25 PGH-2 Tucumcari

Tucumcari had a single T foil at the bow, and main foils aft of amidships, following the general arrangement of High Point. Developments included the aft foils being separate, and having anhedral to minimize tendency for ventilation during turning. Propulsion was by two-hull-mounted water jets taking water from a forward-facing intake at the root of the struts. Water was delivered up the struts to the water-jet pump which ejected the water from exhausts angled downwards through the hull bottom near the stern. Control flaps were fitted to bow and stern foils, allowing the craft to bank in turns, which could be completed in just over 210-m radius at 40 knots. The craft was able to achieve greater than 50 knots in calm water. It had a new control system developed by Boeing incorporating dual-sonar wave sensors, rate and yaw gyro package, and accelerometers, all coupled to a command computer that operates the flaps and steerable bow strut to respond to the helmsman's commands.

Combining the experience from PGH-1 and 2, the US Navy was able to begin selection of the configuration for a class of Patrol Hydrofoils intended to be used as a common patrol craft platform by NATO forces in the Mediterranean. Italy and Germany joined with the USA in this development while other NATO nations took an observer role at the initial stages. In November 1971, the US Navy awarded Boeing a contract for the preliminary design of a 230-t hydrofoil craft, the Patrol Hydrofoil Missile class. Design and procurement were committed for the first two vessels of the class for the US Navy, for delivery in 1975. Following this, it was initially expected that up to 28 such vessels would be built.

Fig. 5.26 PHM hydrofoil fleet in formation

The PHM was significantly larger than Tucumcari, at 133 ft against 75 ft and 230 t against 64 t. A number of simplifications were incorporated. The rear foil system was made as a single unit with a shallow M-formation foil, and vertical struts on the outside of the hull at each side that end in a rotatable connection inside which the waterjet duct feeds water to the two pumps. A single GE LM 2500 gas turbine of 17,000 shp drives the two water jets that exhaust under the hull transom similarly to Tucumcari. The foil control system and flaps follow the same approach as Tucumcari.

The craft have proven successful in their concept, deployed to a forward base, and operated from containerized shore support. The range of these craft from base is designed as greater than 500 nautical miles, running at greater than 40 knots in seas up to 12 ft and a maximum speed of 48 knots. Following completion of PHM-1 by Boeing in 1975, trials and first of class commissioning in 1977, five further craft were built for the US Navy, as follows:

PHM-1 USS Pegasus	July 1977
PHM-3 USS Taurus	October 1981
PHM-4 USS Aquila	January 1982
PHM-5 USS Aries	May 1982
PHM-2 USS Hercules	September 1982
PHM-6 USS Gemini	November 1982

PHM-4 USS Aquila is shown in Fig. 5.26. The leading particulars of the class are listed in Table 5.2. These six craft were operated from a Naval Station at Key West in Florida for coastal patrol and interdiction; see Fig. 5.26. They also carried out extended patrols of up to 10 days. The PHM was able to be refueled from Oiler ships while on Patrol to extend its endurance.

Table 5.2 Example military hydrofoils

Type	PCH-1 High Point	AGEH-1 Plainview	PGH-1 Flagstaff	PGH-2 Tucumcari	FHE-400 HMCS Bras D'Or	PHM-1 Pegasus	Sarancha (Almaz Project 1240 Uragan)
Country	USA	USA	USA	USA	Canada	USA	Former USSR
Builder	Boeing	Grumman	Grumman	Boeing	DeHavilland	Boeing	Almaz shipyard
Mission	Antisubmarine warfare ASW	ASW	Gunboat	Gunboat	ASW	Guided missile	Guided missile
Delivery year	1963	1965	1968	1968	1968	1975	1977
Crew	18	25	13	13	25	24	40
Displacement, t	108–131	328	72	58	236	235	220–330
Loa, m	35.3	64.6	22.2	22.7	49.95	40.5	45
Beam, m	9.53	12.3	6.5	5.9	6.5	8.9	10
Draft, m: hull borne	6.04	7.6	4.26	4.33	7.16	7.1	7.3
Foil borne	2.6	n/a	1.2	1.5	2.3	2.7	2.5
Max speed, knots	50 max 30–40 cruise	50	40	50	63	50	58
Machinery	2×RR	2×GE	RR Tyne	RR Proteus	GT	GT	2×NK12M
Foil borne	Proteus 1273	LM1500 GT	Gas turbine	Gas turbine	Gas turbine	Gas turbine	Gas turbine
kW	2×3,154	2×10,760	2,486	2,450	16,325	15,000	2×11,130
Hull borne	GM 12V71	GM12V71	2×6M6V	Diesel	Diesel	Diesel	–
Power, kW	390	2×390	150	90	1,090	2×440	–
Propulsion	Z drive 5 b 34″ prop	2×Z drive 5 b 40″ prop	Z drive SC prop	Water jet	Z drive	Water jet	Z drive SC prop

(continued)

Table 5.2 (continued)

Type	PCH-1 High Point	AGEH-1 Plainview	PGH-1 Flagstaff	PGH-2 Tucumcari	FHE-400 HMCS Bras D'Or	PHM-1 Pegasus	Sarancha (Almaz Project 1240 Uragan)
Range, nautical miles	n/a	500	800	800	500	600	700
Seakeeping	SS 4–5	SS 5	SS 4	SS 4	SS 4-5	SS 5	SS 5
Main weapon	2 × Mk32 torpedoes	6 × anti submarine torpedo	152 mm M551 howitzer rapid fire gun	Artillery 4 × 40 mm	4 × Anti-sub Torpedo	Artillery 1 × 76 mm 2 × 4 Guide missile "Harpoon"	4 × GM SS-N-9 2 × GM Anti-air, SA-N-4
Foil features	Fully submerged	Fully submerged	Fully submerged	Fully submerged	Surface-piercing foil	Fully submerged	Surface-piercing fwd foil
	Retractable	Automatic control	Retractable	Automatic control		Automatic control	Automatic control
	Automatic control	Retractable foils	Automatic control	Retractable foils		Retractable foils	Retractable foils
Foil loading	68% aft 30% fwd	10% aft 90%fwd	30% aft 70% fwd	70% aft 30% fwd	90% main 10% fwd	68.2% aft 31.8% fwd	30% aft 70% fwd
Frl	1.38	1.02	1.39	1.72	1.53	1.29	1.42

Fig. 5.27 HMS Speedy

In Europe, Italy went ahead to build a single craft, the "Sparviero" (swordfish), in 1974 which was a direct development of Tucumcari, with similar dimensions, foil arrangement, etc. while being fitted with twin missile launchers at the stern and a large gun on the foredeck. The craft was successful enough for a further six of the type to be procured by the Italian Navy up to 1983, and in 1991 the Japanese Maritime Defense force followed up with an order for three craft to the design to be built by Sumitomo Heavy Industries. The design can, therefore, be considered a success.

Other Navies did not follow through on the PHM program. The UK eventually ordered a Jetfoil for fisheries protection evaluation purposes, named HMS Speedy (Fig. 5.27), which was used in a number of trials, but did not lead to a new series construction for the Navy. HMS Speedy was a modified Jetfoil 929-115, built on the commercial Jetfoil construction line at Renton; it was completed in 1980 and delivered to the UK, where it was put into service in June, and completed just less than 2 years' service, being decommissioned in April 1982. The craft was sold to Far East Hydrofoil in 1986, converted to passenger configuration for service between Hong Kong and Macau and has been operating that service since that time. It is interesting that the British Navy did not find success with the Jetfoil.

The capability of the craft was clearly compatible with the North Sea conditions, and the US Navy hydrofoils had shown the way with their operations off the US east coast. At this time, the Royal Navy was still adjusting to a more significant role of economic zone patrol rather than "blue water" long-distance projection of power. The deployment of a single high-speed craft operating only with other rather slower

Fig. 5.28 Tayfun in Baltic

coastal patrol vessels would have given particular challenges. Compared with the available range and endurance of most monohull naval vessels, the hydrofoil was limited and required a different operating philosophy for success. While high speed to catch up with an errant fishing vessel or illegal is important, successful coastal patrol also needs ability to send men aboard the target vessel and this is difficult from a hydrofoil to a monohull, for example. Where the military mission involves search and locate or deployment of weapons such as missiles, the fit is much better.

Developments in Russia

In 1969, an experimental fully submerged hydrofoil, the "Tayfun," was built and tested in the Baltic. It was gas turbine powered, and used the aircraft foil layout with tail foil outboard of the stern supporting a Z drive to the propeller (Fig. 5.28). The craft was arranged as a passenger ferry.

In 1973, the Almaz Shipyard in Leningrad built a large hydrofoil missile patrol craft for the Russian Navy with surface-piercing bow foils and fully submerged stern foil, project 1240, the Uragan (Fig. 5.29). Power for the 320-t displacement craft was from two 15,000-shp gas turbines driving propellers at the rear foil. Top speed achieved was 58 knots, and greater than 45 knots in sea state 5. The craft was on technical trials up until 1977 and was transferred to the Black Sea fleet via the Russian river and canal system in 1979. Operations with the fleet continued until 1992 when she was decommissioned. Just the one ship was built. The armament was heavy and, while capable, the ship was complex to maintain and expensive to operate. Hydrofoils were constructed in titanium, and a radar-based sensing system was used for depth control using flaps fitted to the rear foil.

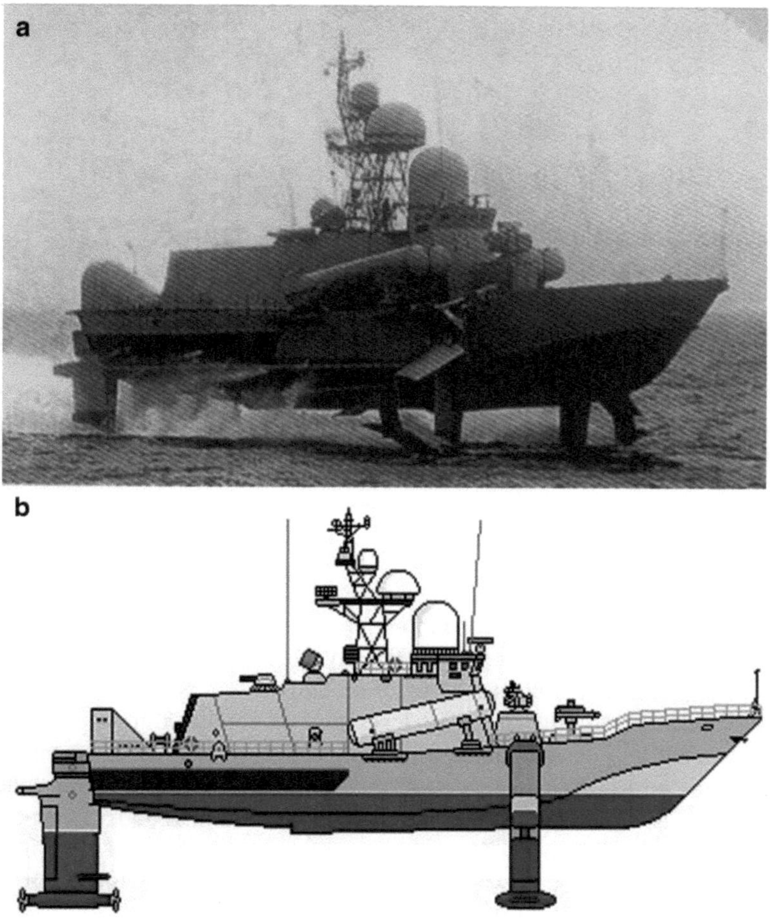

Fig. 5.29 (a) Sarancha class hydrofoil, the Uragan; (b) a side view drawing of the craft

A larger patrol hydrofoil, the 420-t Bobachka, has also been built and commissioned with the Russian Navy in the Black Sea Fleet, the world's largest hydrofoil to date; Fig. 5.30. Again, just a single example has been completed.

Israel

A fleet of three Flagstaff derivatives from Grumman were built at Haifa shipyard, delivered in 1982, 1984, and 1985. The craft were of 105-t displacement compared to Flagstaff's displacement of 67.5 t; the Shimrit class is powered with Alison 501KF turbines rated at 5,390 shp, giving a speed of 48 knots. The craft had a range of 1,850 nm, which was sufficient for its role of protection of the Israeli coastline against potential attack from its neighbors.

Fig. 5.30 Babochka hydrofoil

Passenger Ferry Craft

The next step for Boeing after its military craft was to commercialize the technology developed for PHM, having found the key technical building blocks necessary for success with a fully submerged hydrofoil. The result was the "Jetfoil 929-100," a 110-t displacement craft. The main developments compared to the PHM were a revised planing hull design having wider beam and superstructure giving accommodation for up to 250 passengers in aircraft-style seating layout on two cabin decks. Service speed was set at 42 knots, with a maximum around 50 knots. Power is provided by two Allison 501K gas turbines, each providing 3,300 shp. The water-jet propulsion comprises two Rocketdyne axial pumps, which improved significantly on the performance of the system used for the PHM.

Figure 5.31a shows the principal elements of the Jetfoil propulsion and foil system, i.e., fully submerged canard foil arrangement with a single inverted T strut/foil forward and three struts, full-span foil aft. The forward foil assembly is rotated hydraulically through 7° in either direction for steering. All foils have trailing-edge flaps for controlling pitch, roll, and yaw and for takeoff and landing assistance. Foils and struts retract hydraulically above the waterline, the bow foil forward, and the rear foils aft. Figure 5.31b shows one of the Jetfoil ferries at speed.

The prototype Jetfoil 929-100 was built in 1974 and went through nearly 500 h of testing before delivery to Pacific Sea Transportation Ltd. for service in Hawaii. The first commercial ferry service was begun with craft 002 in April 1975, built for Far East Hydrofoil Company, an operator in Hong Kong. Boeing went on to complete five craft for Hawaii, and additional craft for Far East Hydrofoil and other operators totaling ten craft by May 1977. The 929-100 was improved to become the 929-115 model. This incorporated improved hull design for lighter structural weight, improved front foil design with tapered section, and rear struts

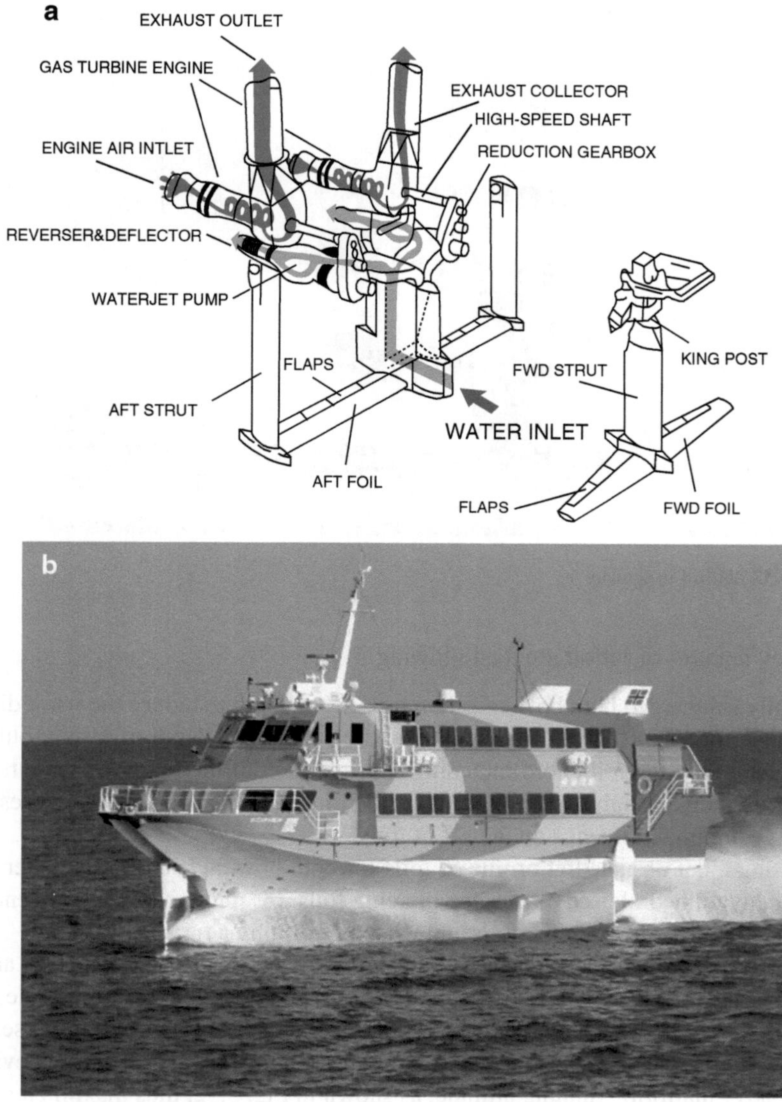

Fig. 5.31 (**a**) Jetfoil propulsion system. (**b**) Jetfoil 929 at speed

with external stiffeners removed for improved hydrodynamic performance. A number of other adjustments were made to comply with international regulations which were brought into place in the late 1970s as fast ferries became more widespread. A total of 17 craft were built, and later modified or updated as they were sold between operators over the years. Boeing stopped production at its Renton, Washington, facility from 1978, but continued to work through its licensee Kawasaki who had built 11 Jetfoils up to 1993.

Fig. 5.32 Jetfoil in seaway

Key features of Jetfoil are the following.

- Fully submerged both bow and stern foils set at depth to reduce the wave disturbance forces in flight: To achieve this, the submerged depth of foils should be greater than the foil chord including consideration of the sea wave height, thus minimizing the submergence effect on the foils, i.e., the encountered waves will have minimal influence on the lift of foils.
- Using a flat or anhedral profile of foils for both bow and stern foils rather than the high dihedral used on surface-piercing foils, for minimizing the influence of encountered waves on the foils and when performing a turn.
- Using an automated control system for controlling the elevators' incidence angle, and thus the lift of both bow and stern foils, so as to give a smooth ride, i.e., obtain small displacement motions and accelerations for vertical, transverse, and longitudinal motions: The craft can be running in rough waves while achieving a steady "platform" running attitude, as shown in Fig. 5.32, thus maximizing passenger and crew comfort, and with small speed loss in waves.
- Using water jet for propulsion, with a forward-facing water inlet at the stern foil strut and the pump impellor at the hull elevation, resulting in no complicated power transmission system on the craft caused by the deep-submerged foil system.
- Since the foils create a deep draft for the hydrofoil, such craft can only operate at deepwater jetties unless the foils are made retractable. This has been arranged on the jetfoil, with the water jet having an alternative inlet at the hull underside for slow-speed maneuvering.

The Jetfoil has a ride that is almost aircraft like. Quoted vertical acceleration is 0.04 g in a 2-m sea state. The craft really is amazingly steady and smooth, as can be confirmed by the authors from riding on a jetfoil. It has had great success on certain

Fig. 5.33 Rodriquez FSH-38

routes, the Hong Kong to Macau route being a key operation, while on others the success was rather more limited, mainly where an operator had what seemed like a good idea, but the clientele did not materialize. One such was a service across the channel between Dover and Zeebrugge in Belgium. Unfortunately, there was not the same passenger density available on this route as there was on Dover to Calais; a route that is often used by English holidaymakers for day trips to France. There might have been greater success if the craft could carry cars, as did the slower conventional ferries, though perhaps not, as it is easier for people to go to Calais and drive up the coast to Belgium.

Once Boeing had stopped its own developments from 1985, the US Navy having come to a halt with fast craft, and the Jetfoil having reached its potential market, hydrofoil advancement for passenger ferries also came to a halt in the USA, and while the fleet continued with reliable services throughout their operational life, the future began to look more like catamarans, or perhaps a catamaran with submerged foils like the Kvaerner Fjellstrand Foilcat below or maybe simply foil assistance as we discuss in Chap. 7.

Italy

Since 2004, Rodriquez in Italy has been developing its own fully submerged hydrofoil passenger craft designated the FSH-38; Fig. 5.33. Two craft have been built, funded jointly by the European Union and Italy's Ministry of Education and Research. The first has a traditional power train from Rodriguez with centrally located engines and canted drive shafts down to stern-mounted propellers. The second craft has a Z-drive propulsion with a pair of contra-rotating puller propellers at the base of the propulsion strut [5-8].

The foil arrangement is of canard form with a forward Tee foil and aft main foil supported by two vertical struts. All the foils have flaps and rudder surfaces. Rodriquez has chosen to separate the power system from the rear foils. This allows the engines to be installed amidships making weight distribution easier for a diesel-powered craft. The canted shaft propeller drive is from vee boxes forward of the engines, minimizing the shaft angle. The engine takes its cooling water from a snorkel intake on another strut for the vee drive craft while for the Z-drive version this intake is incorporated in the propeller strut.

The hydrofoils have been the main subject of the R&D project, investigating the hydrodynamics of the foils and structural loads in waves. Key characteristics of this craft are, LOA 37.25 m, BOA 8 m, displacement 140 t, 243 passengers. Power is from 2 × MTU 16V4000M70 diesels, providing a total of 4,640 kW giving a planned service speed of 45 knots.

Trials with the canted propeller drive version from late 2009 have verified the performance in a seaway, with accelerations as low as 0.07 g RMS in 1.5-m seas which are typical of the service conditions in eastern Mediterranean. The diesels give a useful fuel consumption of around 1,000 L/h, which is between 10 and 20% better than the existing fast ferries. Rodriquez is, thus, going back to its roots aiming to provide a passenger ferry with high earning power for the operator, and providing a route for upgrade for its surface-piercing hydrofoil clients.

This craft represents a new step in submerged foil craft design, since diesel power is rather more fuel efficient and less demanding form the maintenance point of view, and so should result in a revived market for such craft. Electronics and control systems have advanced since the 1990s so that this aspect of design is not the core challenge any more. Hydrofoil dynamics for submerged foils operating in the 40–50-knots range are understood now, so optimization for a mission requirement is realistic. The interesting aspects of Rodriquez' FSH-38 are the power system, and whether the Z-drive system can give improvements over the traditional long shaft drive. Further trials of the second prototype in 2010/2011 and early operator experience will test this fully.

China

In China, the technology of fully submerged foils was studied through the 1980s as the Jetfoil was refined by Boeing and Kawasaki, and this research enabled the design and construction of the PS-30, a ferry based on deep-submerged hydrofoil technology. The "PS-30" was built in 1995, with displacement 118 t, max length 29.1 m, max breadth 9.2 m, molded depth 2.6 m, draft 4.51 m, passengers 294, crew 9, service speed 43 knots, all-aluminum-welded hull, and two Allison-501KF gas turbines as the main power plant, with output of 3184 kW each, driving two water jets for propulsion. The craft has been operated in the Pearl River delta to Hong Kong since its delivery in the mid 1990s. Figure 5.34 shows the PS-30 in service.

Fig. 5.34 Chinese PS-30

One can see the water-jet inlet on the struts of rear foils, outlet nozzle at the stern, and the gas turbine driving the pump impeller of water jet system, via the gear box. The flaps on each foil are controlled by an automated system via hydraulic power for trimming during voyages, and dynamic adjustment for improved craft motions in waves. Figure 5.35a–f shows details of the PS-30 design, including the power plant, foil configuration and water jet pump impellor.

Norway

In Norway, during 1989, both Westamarin and Fjellstrand, successful builders of catamaran ferries (see Chap. 6), designed new versions of their catamarans with submerged foils aimed at achieving higher service speeds. The Westamarin design did not go further than the drawing board while Fjellstrand first built a small-scale test craft and then a full-scale prototype ferry, the 40-m Foilcat in 1994, based on its normal ferries, but with gas turbine-driven water-jet propulsion with two intakes at the outer base of the aft foil spanning between the twin hulls and two forward foils on struts close to the bows of each hull, with hydraulic incidence control for depth

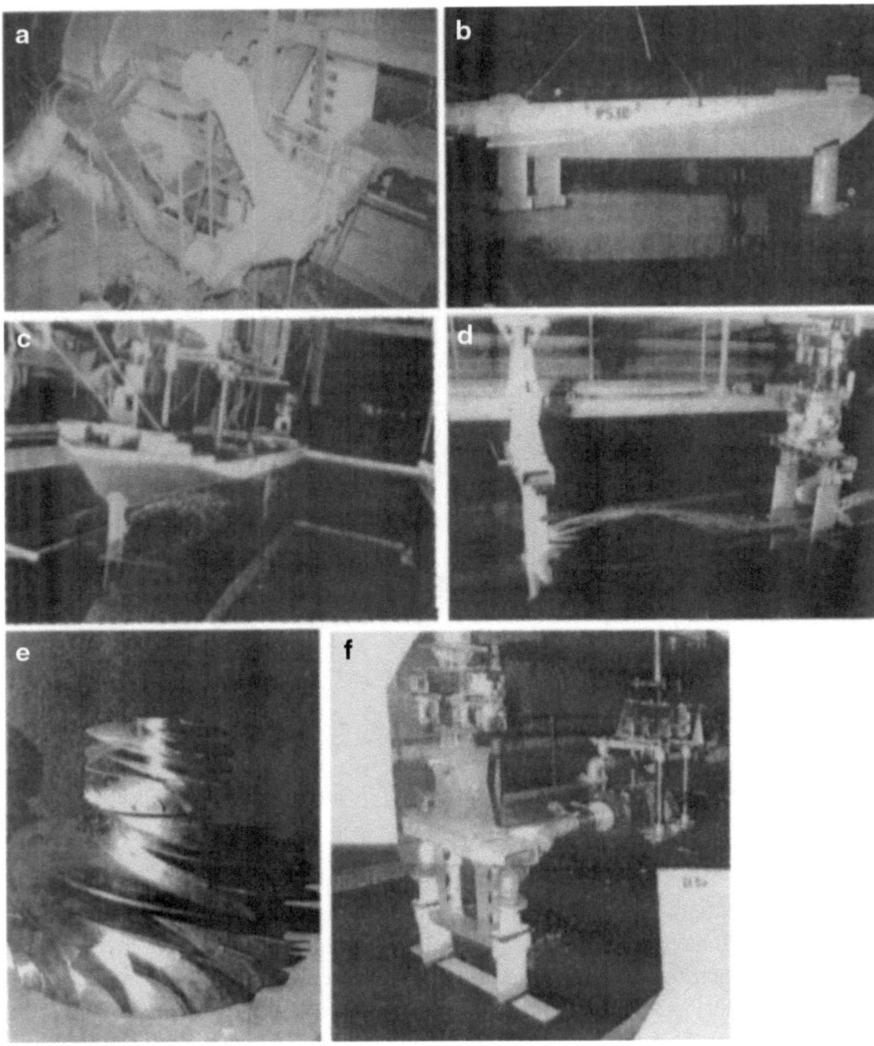

Fig. 5.35 Details from PS-30: (a) Gas turbine power plant in the engine bay; (b) experimental model; (c) model testing in towing tank, frontal view; (d) model testing in towing tank, profile view; one can see the influence of wake causing by fore foil to rear foil; (e) pump impeller of water-jet installation; (f) water-jet propulsion system combination diagram

control in flight; Fig. 5.36. The flight height was arranged for the catamaran hulls to be just clear of the water, rather than the higher flight of the Jetfoil.

The prototype Foilcat did not reach its target speed, and after evaluation Fjellstrand concluded that the solution would be to shorten the craft to 35 m to reduce weight. The self-funded prototype was, therefore, converted back to normal ferry configuration and sold for operation in Bulgaria. Meanwhile, Far East Hydrofoil ordered the first two operational Foilcats and placed them in service

Fig. 5.36 Foilcat Penha in service between Hong Kong and Macau

between Hong Kong and Macau from 1995. The earlier prototype problems appeared to have been solved and introduced a craft with spacious accommodation for 420 passengers at 50-knots service speed, similar to that of Jetfoil. Since this time, there have been no other sales for Foilcat, so it is clear that while a technical success, it is only the high-density passenger route in Hong Kong that can support the economics of such high-speed hydrofoils.

Switzerland

Since December 1997, Supramar has collaborated with the SEABUS-HYDAER Consortium, a group of 13 specialized European companies focusing on the development of new solutions to eliminate cavitation/supercavitation at very high speeds above 100 kts. Funding has been provided from the European Commission in Brussels under Brite Euram. Supramar's task has been the development and layout of the hydrofoils for a hydrofoil-assisted wing in ground-effect fast ferry with a total displacement of 500 t, 800 passengers, and 100 cars, and a range of 850 km with a speed of 120 kts. The work of Supramar within this Brite Euram project was backed by a grant of office fédéral de l'èducation et de la science (OFES), which financed the tests in the high-speed cavitation tunnel of laboratoire des machines hydrauliques (LMH) at the école polytechnique fédérale de Lausanne (EPFL) in Lausanne Switzerland [5-9,5-10]. The results have been some useful developments in cavitation inception and boundary layer control which should allow optimization of future hydrofoil craft.

Moving On

Now that Rodriquez has completed development of its fully submerged foil craft, we have the prospect of a resurgence in hydrofoil passenger ferry services through the upgrading from currently used craft. While the catamaran has developed many routes where large passenger numbers are available to finance the larger craft, there are still many smaller communities for which a high-speed marine link can improve logistics; so there continues to be a market for such craft in the Mediterranean and on many coastal routes, where a craft such as the FSH would be a useful upgrade.

Deep-submerged hydrofoil craft can be characterized as follows.

Advantages

- *Seaworthiness*: These craft have very small motions and so are comfortable for passengers and crews, causing less seasickness and small speed loss in rough waves.
- *Maneuverability*: By using controllable flaps on foils at both sides of the craft, operators can use a banked turn to obtain inward heeling angle and achieve small turning circle and good maneuverability as shown in Fig. 5.32.

Disadvantages

- *Deep draft and complicated retractable foil system*; in addition, in the case of retracting the foils, the transverse stability is reduced while boating.
- *No inherent stability at service speed*, so the automatic control system has to be highly reliable, consequently, with high cost. Without pilot actively controlling the craft in flight, this type of craft would eventually come to grief, in the same way as a jet aircraft.

Finally, before we turn to catamarans, a reminder of the basic options for hydrofoils and issues that a designer has to take into account:

- Foil plan form, load distribution, and foil geometry
- Arrangements for takeoff assistance
- Arrangements for roll, pitch, and heave stabilization (static or dynamic)
- Propulsion arrangements foil and hull borne
- Momentum drag of water jets
- Cavitation of foils
- Cavitation of propellers (and indeed of water-jet pumps)
- Foil retraction for boating
- Foil retraction or other means for cleaning and maintenance
- Water depth at terminals

Chapter 6
Catamarans and Multihull Craft

Canoes and Outriggers

The history of the catamaran goes back almost to the time when man first used a tree trunk for overwater transportation. While conventional monohull ships had their beginnings in the dugout canoe or pirogue, the catamaran's origin was the raft formed by lashing two or more logs together.

The traditional pirogue, Fig. 6.1, while efficient to paddle, is not very seaworthy. It is OK in a river where waves are small, but at an estuary or coastline something sturdier is needed. Lashing two craft together, perhaps directly, or with a little space between creates a wholly different transverse stability, enables much stronger paddling, and with a mast rigged can take a sail. So by creating a catamaran configuration we gain:

- Improved transverse stability, dependent on the space between the two boats
- A more spacious platform for crew and cargo
- Improved performance in waves

The Polynesians are credited with constructing the first seaworthy ocean-going catamarans, at least a thousand years ago. They brought this craft to such a high state of development that they were able to make voyages of exploration over vast expanses of the Pacific, from Tahiti to Hawaii, Easter Island, and New Zealand. Figure 6.2 shows a typical Polynesian proa catamaran. The craft were fast, and could out-manoeuvre the ships of Cook's fleet on his voyage of discovery to the Pacific in the eighteenth Century.

Through the sailing boat era, sail-carrying ability was the most important factor for obtaining performance. The optimum hull form which will sail fast given some forward thrust from the sails unfortunately has the least stability to stand up against the heeling moment of the sails when the wind is from one or other side of the ship.

L. Yun and A. Bliault, *High Performance Marine Vessels*,
DOI 10.1007/978-1-4614-0869-7_6, © Springer Science+Business Media, LLC 2012

Fig. 6.1 (**a**) Pirogues in Gabon. (**b**) Pirogues in Gabon 2009

In European ships, this stability was largely achieved by carrying ballast in the form of stones in the bilge spaces either side of the ships keel and by extending the keel to greater depth. This, however, has the penalty of increasing displacement and thus requiring even higher amounts of sail to be carried to achieve the same speed.

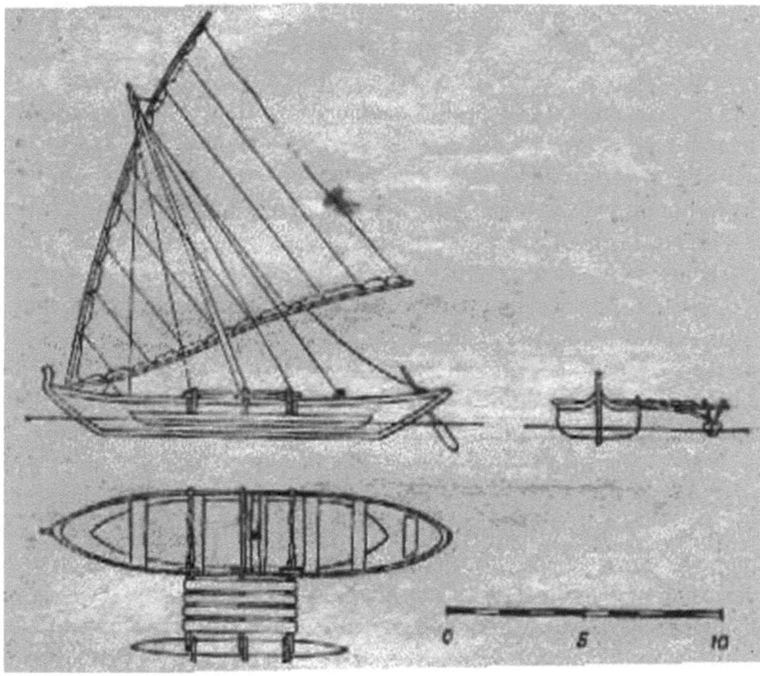

Fig. 6.2 Polynesian Proa

A catamaran is able to carry the large amount of the sail necessary for high speed by countering heeling moment with the inherent stability of two widely separated hulls, which may then have a fine (thin) hull form. In recent times the racing catamaran, trimaran, and both of these vessel types with hydrofoils have all been developed to a state of optimization with lateen sail plans that enable them to reach 40 knots and sail up wind as high as 30° to the oncoming wind.

Development of the mechanically powered high-speed catamaran, i.e. with Fr_L number around 0.7, can be traced to the Second World War. During that time, the combination of materials, engines, and equipment necessary for a high-speed craft became available due to the construction of parts for aircraft that were built in large numbers as fighters and bombers. Structural material such as high strength aluminium alloy and steel, powerful high-speed diesel engines for propulsion, and hydraulic equipment for power transmission and ship outfit became available as fallout from the wartime design and construction of fighter planes, tanks, and patrol boats. Taking advantage of these advanced materials and engines, the speed of monohull planing boats (torpedo and patrol boats), was able to reach 40–50 knots. The heaving and pitching motions, accelerations, and impact or slamming loads in short seas were so large on these craft that it was necessary to decrease the engine power and reduce speed in any significant seaway. This low seaworthiness meant that operations were quite weather limited in open ocean conditions, a disadvantage for military craft.

Aiming to reach higher service speed in a seaway and also minimize propulsion engine power, the hydrofoil craft, air cushion vehicle, surface effect ship, and WIG craft were all developed step by step in the 1950s–1980s and began to demonstrate the benefits of reducing drag forces by lifting the hull out of the water. However, due to the complex outfit needed for these craft types—skirt and lift fan systems for ACV and SES, sophisticated hydrofoil geometries and automated control systems for hydrofoil craft and the aircraft quality technology for WIG—the construction cost for such craft is high and maintenance can be a burden. Many early operators of these special craft were not familiar with the high level of maintenance needed in order to retain performance, and so cancelled ferry crossings were more than a rare occurrence, leading to loss of confidence by the customer. Not a solid basis for good business. All these reasons combined to limit the development of ACV/SES/HYC, particularly for civil applications, since without customer and operator confidence the market could not grow.

In the 1980s, naval architects found that high-speed conventional ships with very high length/beam ratio (slenderness) to reduce wave-making resistance at high speed (Froude Number Fr_L, higher than 0.7) tended to have low transverse stability unless the traditional approach of adding ballast were used so as to increase the metacentric height. Permanent ballast is unusable payload and so reduces a vessel's economy, whether by reducing income generating payload, or reducing the fuel that can be carried and hence the vessel endurance. The better option perhaps was to change a monohull into a catamaran with a space between two hulls, so as to form a craft with less resistance, high transverse stability, etc., while retaining as far as possible simple construction and equipment installation, retaining low cost. This was the logic which led to Westamarin's catamaran passenger ferries of the 1970s after they had built a number of hydrofoils for service along the Norwegian coast [5-6] and also influenced the development of catamaran ferries in the Puget Sound [6-1] and the continuing application of catamaran ferries in South Western Norway to the present [6-2].

The catamaran has developed vigorously since the 1980s, particularly in civil applications, and really was a new force coming to the fore in the HPMV family. Almost 60% of the high-speed ferry market is currently supplied by catamarans of various designs worldwide! Tables 8.1 and 8.2 in Chap. 8 show HPMV new buildings in recent years, The tables show that catamarans have occupied the majority of the HPMV market over the last 20 years or so as they have developed in size from passenger only craft up to large passenger and cargo vessels.

Figure 6.3a shows a typical catamaran with asymmetric transverse section configuration of the two hulls, as first introduced in Norway by Westermoen in the 1970s. This configuration looks like a conventional ship has been cut longitudinally into two parts along the centre line, and moving them apart by a distance. Since then several different hull configurations have been developed in an attempt to optimize resistance and motion response in a seaway. We introduce these below.

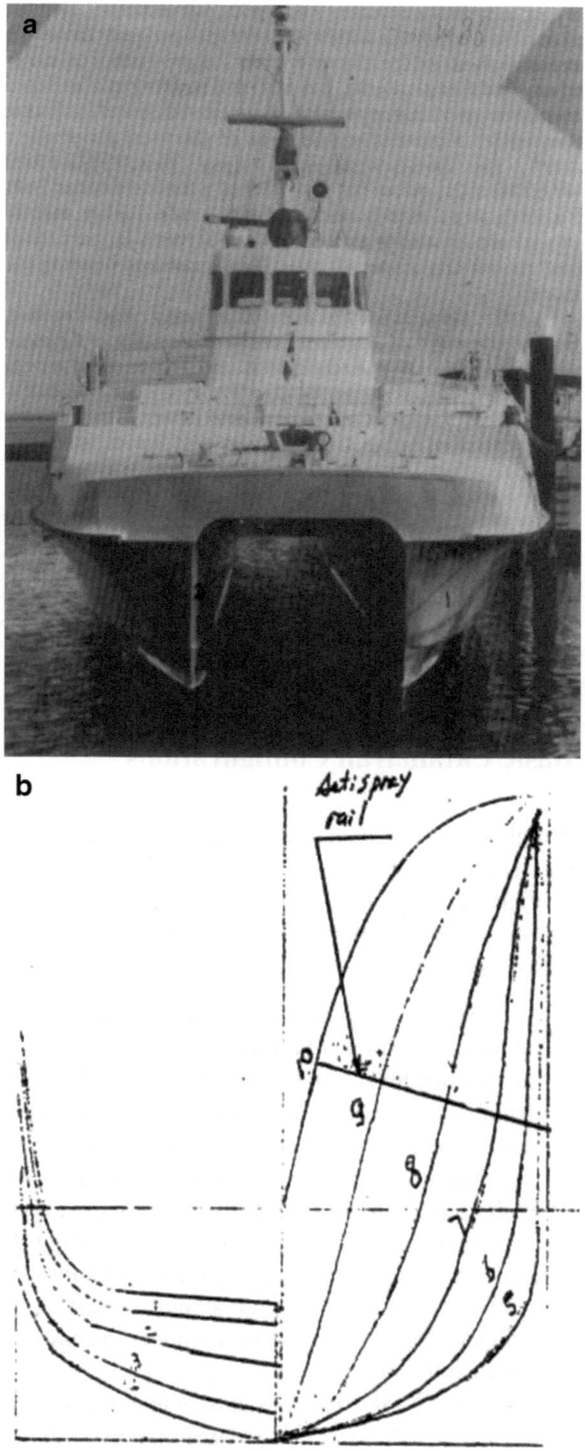

Fig. 6.3 (**a**) Asymmetric hulls. (**b**) Symmetric round bilge hulls. (**c**) Symmetric planing hulls

Fig. 6.3 (continued)

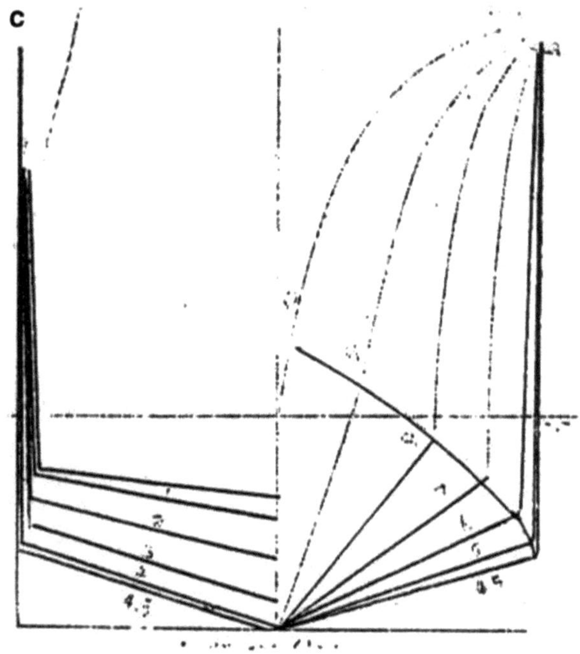

Basic Catamaran Configurations

A high-speed catamaran generally has a slender hull form. It may be expected that high speed may generate planing lift; however, such a slender body is difficult to be configured as a planing surface compared with the lower L/B of a monohull, so the body line configurations of the catamaran hull are generally designed as three basic types, as below:

1. *Asymmetric body plan*, shown in Fig. 6.3a, in which there is a vertical wall on the inboard side of each hull, to decrease unfavourable wave interference between the two side hulls and thus reduce wave-making resistance. This is important particularly in case of a catamaran with small separation between the side hulls, and at low speed, where the effect of unfavourable interference from wave-making is significant. Another option is to use an asymmetric body plan with straight side of the body plan outboard, to reduce the wave-making caused externally that affects piers and river banks, for craft operating in rivers or narrow channels.
2. *Symmetric round bilge configuration*: similar to the lines of high speed displacement conventional monohull ships, i.e. destroyers at Fr_L around 0.5, while the catamaran L/B is much finer. Figure 6.3b shows a typical round bilge type catamaran hull body plan.
3. *Symmetric hard chine lines* aims at high speed service where hard chine lines and full shape at the stern can provide high hydrodynamic lift at speed, adding to the

buoyancy, and also plenty of space for the main engines and water jet propulsion installations as well as the water inlet and steering devices. Figure 6.3c shows a typical body plan of a hard chine type catamaran hull providing more hydrodynamic lift at higher speed.

Designers make their choice between configurations based on the catamaran mission, using straight sides whether inboard or outboard where the reduced wave-making is required to boost craft performance, or reduce interaction with environment. In addition, the designer can use differing height and geometry of the lower surface of the bridging structure to create additional aerodynamic lift from the tunnel roof.

Features of High Speed Catamarans

So what are the key features of a high speed catamaran, in comparison to other HPMV, particularly monohull craft? We outline the main ones below. Leading particulars of some example craft are listed in Table 6.1 at the end of this chapter to give a feel for the data.

- *Low wave-making resistance* due to high length/beam ratio allows the designer to minimize the propulsion engine power rating for a given service speed. The slender hull form does result in increased wetted area, about 40% larger than that of conventional monohull ship, and thus increasing friction resistance, which is dominant at low craft speeds. Powering for acceleration through hump has to be carefully considered therefore, so as to ensure adequate acceleration to service speed in higher sea conditions.

Optimum relative speed (Froude Number, Frl) for a displacement catamaran in general is in the region Frl = 0.6–0.95, i.e. about 23–35 knots for a catamaran with length of 35 m. As size increases, the optimum speed will also increase; for example, a catamaran with a length of 110 m the optimum speed will be as high as 50–58 knots!

- *High transverse stability* due to the space between hulls, and in general, the transverse metacentric height, GM, will be about ten times higher than a monohull ship

Large deck area, due to the centre bridge between the side hulls this is generally greater than monohull ships by 40–50%, thus giving spacious and comfortable passenger cabins and other working cabins. Figure 6.4 shows the general arrangement of a typical modern passenger catamaran, in which two wide passenger cabins with aviation type seats are arranged on the upper deck and main deck. Twin high-speed diesels are located in the symmetrical geometry side hulls, driving two sets of open propellers at the stern. The bridge is arranged above the passenger cabin to give wide vision for the crew.

Table 6.1 Leading particulars of example Catamarans (early)

Name	W86	W95	W100	W3700	31.5 m	38.8 m	Marinjet 41CPV-SD	Marinjet 33CPV
Builder	Westa-marin	Westa-marin	Westa-marin	Westa-marin	Fjellstrand	Fjellstrand	Marin-teknik	Marin-teknik
Country	Norway	Norway	Norway	Norway	Norway	Norway	Sweden	Sweden
Year	1971	1974	1980	1988	1981	1986	1985	1988
Operation	Norway	Norway	Norway	Norway	Norway	Norway	Sweden	Sweden
Passenger (deck level)	174 main	248 main	300 main	322 main	291 main	449 main	306 main	256 main
Loa, m	26.7	29.1	31.9	37.0	31.5	38.8	41.5	33
Boa, m	9.0	9.0	9.8	9.5	9.4	9.4	11.0	9.4
deck area, m^2	184	243	278	310	267	328	410	279
Draft, m	1.2	2.2	2.0	2.0	2.05	1.58	1.2	1.2
Engines	2 MTU	2 MTU 12V396 TB93	2 MTU 12V396 TB93	2 MTU 16V396 TB94	2 MTU 16V396 TB83	2 MTU 16V396 TB83	2 MTU 16V396 TB84	2 MTU 12V396TB83
Power, kW	2×815	2×1,150	2×1,323	2×2,040	2×1,500	2×1,510	2×1,950	2×1,225
Propulsion	2 prop, 3 b	2 prop, 3 b Servogear	2 prop, 3 b Servogear	2 Waterjets KMW S63	2 prop	2 Waterjets KMW 63S	2 Waterjets MJP 650	2 Waterjets KMW S63
Speed light, kt	28	32	30	35	30	37	42	35
Speed loaded, kt	26	30	28	32	28	32	38	32
Frl	0.89	0.97	0.87	0.95	0.88	0.98	1.07	1.00

Name	CP20 Blue Hawk	CP30 III Coral	Islander	Our Lady Patricia	Liuhua Hu	Green Island express	Capricorn	Cat for China
Builder	Mitsui	Mitsui	Incat Tasmania	Incat Tasmania	A Fai	NQEA	NQEA	Austal
Country	Japan	Japan	Australia	Australia	Hong Kong	Australia	Australia	Australia
Year	1975	1988	1982	1986	1984	1982	1983	1990
Operation	Japan	Japan	Australia	UK	HongKong	Australia	Australia	China
Crew + passenger, deck level	5 + 180	5 + 250	118	452	150	230	326	309 main 120 upper
Loa, m	26.5	40.9	20	29.6	21.9	23	29.2	36
Boa, m	8.8	10.8	8.7	11.2	8.7	8.7	11.2	11.5
Deck area, m^2	209	397.5	161	300	169.6	180	300	350
Hull depth, m	n/a	n/a	2.42	n/a	2.7	2.7	n/a	3.6
Draft, m	1.18	1.3	1.6	2.2	1.6	1.7	1.76	1.3
Engines	2 MTU, 12V 331 TC82	2 Fuji 12PA4V 200VGA	Detroit diesel	2 MTU 16V396 TB93	Isotta	2 GM12V 92TA	2 GM16V 92TA	2 MTU, 16V396, TB83
Power, kW	2 × 920	2 × 1,950	2 × 404	2 × 1,430	2 × 551	2 × 594	2 × 890	2 × 1,470
Propulsion	2 prop	2 prop	2 prop	2 prop 5b fixed pitch	2 prop	2 prop 5b albronze	2 prop 5b albronze	2 × MJP, J650R, Waterjets
Speed light, kt	28.5	35	28	31	29	29	29	35
Speed loaded, kt	25	32	24	29	25	25	26	30
Frl light	0.91	0.90	1.03	0.94	1.02	0.99	0.88	0.96

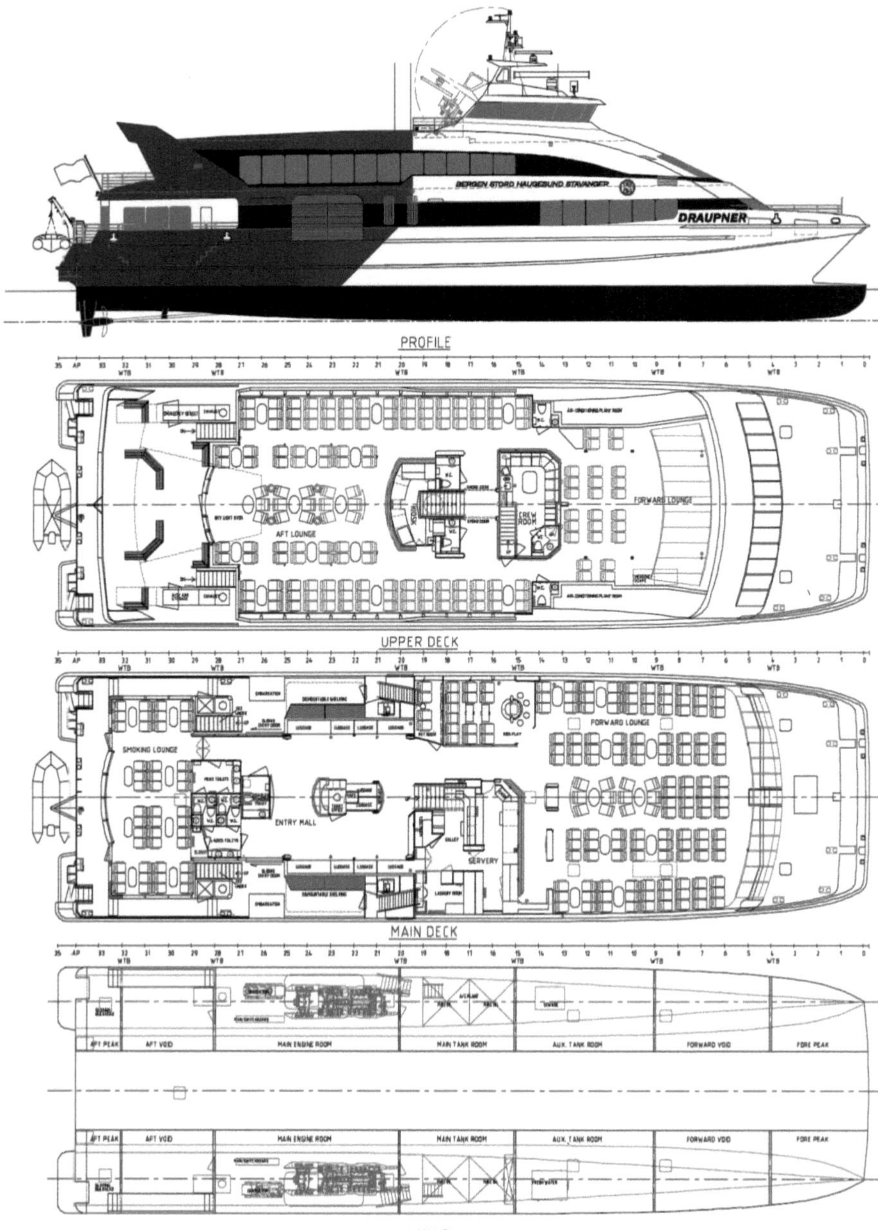

Fig. 6.4 General arrangement of a catamaran Passenger Ferry, Austal 42 m delivered 1999

High manoeuvrability and course stability due to the space between twin propulsors (compared with a monohull) giving a larger turning moment. In addition, due to the stiff transverse stability, captains can use larger rudder angle without heeling, see Fig. 6.5.

Fig. 6.5 "HSV" Swift turning at speed without heel

Low impact and slamming loads as well as speed loss in waves due to hull slenderness compared with monohull craft.

Subdivision against flooding. The catamaran provides a high safety level against hull damage, due to many bulkheads in both side hulls with small individual compartment volume, and thus smaller flooding in case of damage.

Ability to scale up: as the configuration and outfit of a catamaran is relatively simple, there have been no technical obstacles to develop the catamarans from displacements of 100 t, length 20 m, speed 35 knots, up to 3,000 t, 120 m, speed 50 knots in only 25 years since the 1980s.

Payload capacity sufficient to install high-speed diesel propulsion. A key driver for market growth in catamarans has been their use of diesel power rather than the gas turbines used in faster monohull craft, and amphibious hovercraft ferries. While the SES also developed due to using lower cost diesels (initial cost, maintenance, and fuel costs), the catamaran was able to take this up to a much large scale. The Stena HSS were an exception to this, using gas turbine power, and eventually fuel costs did impact on their economy.

Key challenges for development of the high speed catamaran in its earlier forms were as follows:

Heavy Hull Weight

Compared to an equivalent displacement monohull a catamaran has higher hull mass due to the additional material forming the inner walls of the two side hulls, and the deck bridging structure, so the craft could only be used for payloads with load density, such as passengers and/or car–passenger ferry craft. The high hull weight also initially limited the speeds that were practical to obtain for a given installed power. The advantage over monohull ferries was therefore limited in the beginning.

Poor Comfort in Quartering and Beam Seas

The catamaran configuration is a double edged sward for the designer. On the one side, fine transverse stability and low speed loss in a seaway are both basic attributes of any catamaran; however, on the other side due to the strong transverse stability because of the space between the hulls, and decreased longitudinal stiffness because of slender twin hulls, the catamaran profile rolling and pitching natural periods are very similar. Consequently the coupled rolling and pitching motion causes significant discomfort for the passengers and crew, particularly in the case of running in quartering seas.

Performance in a seaway has been a significant challenge for high-speed catamaran development [6-3]. The pitch and roll natural period issue is difficult to change without generating an inefficient hull form, so attention has turned to changing the damping coefficient of the hulls in roll and pitch. To increase damping automated pitch control foils and transverse anti-rolling fins have become widely used on modern high-speed catamaran and do improve high-speed catamaran seaworthiness significantly.

The hull weight issue has been addressed by high-speed ferry shipyards by turning to welded aluminium as the material of choice for the hull structure for larger craft; and the motion response has been tackled by using a semi swath hull transverse section on craft such as the Stena HSS, and many different wave piercing catamarans (WPS). Small catamarans which have to plane to achieve desired speed above 25 knots have turned to fibre reinforced resin construction. We give data from some recent craft designs in Table 6.2.

Let us look at the range of missions currently part of the high-speed catamaran portfolio. These missions have led to evolution of the initial displacement hull form towards the semi-planing form, as well as the slender water plane form used by Stena. We introduce a little later the extension of this to become the wave piercing form, after discussing the planing hull form for catamarans.

First the main missions that have driven catamaran development:

Passenger and Car/Passenger Ferry Craft

Due to its high speed and low construction and maintenance costs, the high-speed catamaran has become popular with many short distance passenger ferry operators worldwide. The market share in the high speed craft community has stabilized at about 65–70% of the total market. Figure 6.6 shows a high-speed catamaran ferry

Table 6.2 Leading particulars of example CFRP passenger catamarans

Name	Sognekongen	FoldaFjord	Krilo II
Design	Brodrene Aa 38	Brodrene AA 25	Brodrene Aa 40
Builder	Brodrene Aa	Brodrene Aa	Brodrene Aa
Country	Norway	Norway	Norway
Year	2010	2010	2011
Operation	Norway	Norway	Croatia
Application	Ferry	Ferry	Ferry
Passengers	295	97+3	351
Cars	–	–	–
Displacement, t	480	199	500
Loa, m	37.5	24.75	40
Boa, m	10.8	8	10.8
Hull depth, m	3.0	2.6	3.0
Draft, m	0.9	0.95	0.9
Speed, knots	34	25 (30 max)	33
Main engine	2×MTU, 16V2000M72	2×MAN, D2842LE410	2×MTU, 16V2000M72
Power, kW	2×1,440	2×749	2×1,440
Propellers	2 KMW 56A3 WJ	2 Servogear CPP	2 Servogear CPP
Stability fins	None	None	None
Structure	CFRP	CFRP	CFRP
Frl	0.91	0.82	0.86

Fig. 6.6 Chinese 42 m cat

constructed in China in 1996, with length 42 m, displacement 158 t, speed 43.5 knots, 380 passengers.

Following the popularity of passenger ferries, catamarans were scaled up to the size necessary for car and truck transport as well as passengers. The breakthrough happened primarily in Australia in the 1980s. Scandinavian builders such as Westamarin in Kristiansand did not have large enough facilities to build craft of this size so as to compete immediately, and the Australian yards found they had a cost advantage while having the technical capability to construct the aluminium hulls necessary for economic structure weight to displacement ratio.

Jumbo Car–Passenger Ferry

At the end of the 1980s, the ferry operator Stena had an opportunity to introduce fast craft on its services between UK and Europe as well as Ireland. They teamed with shipbuilder Finnyards to develop the HSS (High-Speed Ship) and expand the speed and size envelope of these craft significantly. Figure 6.7 shows the HSS 1500 Stena Explorer, constructed at Finnyards, with length 120 m, beam 40 m, 1,500 passengers, 375 cars, or 50 trucks plus 100 cars, service speed 40 knots, welded aluminium hull structure, 4 gas turbines type LM1600 and LM 2500 as main power plant, and driving 4 water jet propulsors.

Figure 6.8 shows the HSS 1500 general arrangement. One can see from the cross sections that the waterline is restricted displacing volume under the water surface, thus increasing the damping coefficient of rolling and pitching, and decreasing the moment of inertia of the water plane, thus increasing both rolling and pitching inherent period, particularly extending the difference of static pitching and rolling natural periods and extending that with increasing speed, consequently improving the seaworthiness of the ship.

Figure 6.9 shows a comparison of pitch periods (left) and roll period (right) resulting from decay tests conducted by Stena Rederi of conventional catamaran hulls, HSS760 and HSS1500 [6-4].

The Stena HSS service from Hoek van Holland to Harwich was a most efficient service between Holland and the UK while the two craft were in operation in the 1990s. Author Bliault took many trips to UK with these ships while working in The Hague and found the ride very smooth in the choppy waters of the southern North Sea.

The ability to get across to UK with a car within 4 hours is in marked contrast to the situation once again, where the same trip is an overnight affair with a

Fig. 6.7 HSS 1500 Stena explorer

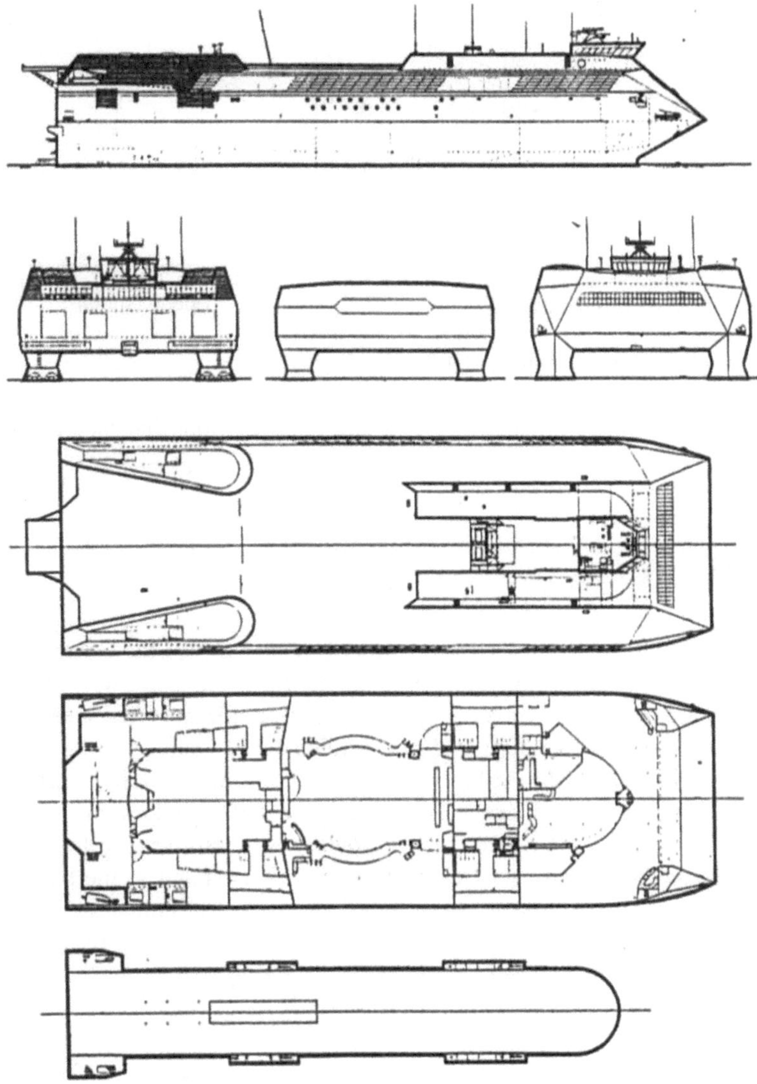

Fig. 6.8 General arrangement HSS 1500

conventional super ferry. The HSS maintained the service at 40 knots and had very few cancellations in winter. The trip time was just sufficient for a passenger to take a meal, and to watch a film in one of the two small theatres on board. Motions did not intrude apart from exceptional circumstances. Unfortunately, fuel costs have accelerated in the last few years and so with their gas turbine main power plant the HSS became marginally economic. While the smaller HSS still operates to Ireland from UK, one of the HSS 1500 craft has now been transferred to Venezuela where operations are less affected by fuel cost.

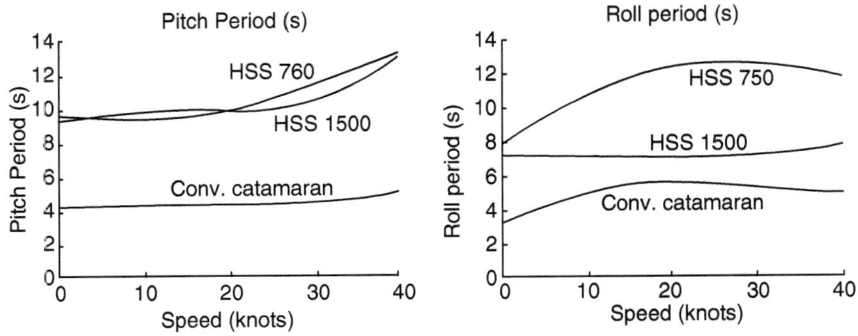

Fig. 6.9 Pitch period comparison catamarans

Turning to the military market, we have two main missions that catamarans have been applied to so far:

Naval Patrol Craft

Figure 6.10 shows the Rodriquez built Haras 1 WPS built for the Oman Navy for coastal patrol duties. This catamaran is 52 m length, 15 m beam, with 10,000 kW power supplied by 4 diesel driven water jet propulsors for a service speed of up to 40 knots and is one of two delivered to the Sultanate of Oman in 2008 and 2009. This craft was designed by AMD Marine Consulting of Australia together with Rodriquez. Three other craft are to be delivered in 2010 and 2011 for ferry duties between the mainland and islands off the South East coastline. The Indian Ocean has a significant wave and swell environment in this area and so the high quality seakeeping of the wave piercing form and forward central hull were key to this delivery. The craft also have a T form stabilizer foil mounted under the central hull almost directly under the wheelhouse location.

Logistic Support Vessel

Shown in Fig. 6.11 is the Westpac, with length of 101 m, 26.7 m beam, with 28,800 kW power provided by 4 diesel driven water jet propulsors for a service speed of up to 36 knots operating in the USA. This craft was purchased from Austal Australia in 2001 to test its capabilities in service prior to the US Department of Defence taking the decision to procure a series of vessels for logistic support across the Atlantic and Pacific oceans for a joint US Navy and US Army programme. The alternate for this programme was the Incat Wavepiercer—the HSV-2 which was also purchased for trials and has seen much active across several oceans while it has been in service. Both Craft have been operated by the US Marine Corps in service.

Fig. 6.10 Haras 1 Patrol Catamaran

Fig. 6.11 Westpac

The JHSV programme is aimed at fast transfer of 550 tonnes or more of troops, military vehicles and equipment "between operations theatres" at speeds averaging 35 knots. The endurance has been set at 1,200 nautical miles which is sufficient minimize interference by refuelling at sea or intermediate locations for most deployments. The vessels will be non combatant vessels and so the hardening and redundancy of systems needed for front line units is not needed for the HSV which can be built essentially to commercial specifications. Command and control systems will nevertheless be to military standards to allow helicopter operations on a 24 h basis.

The crew will be between 22 and 40 dependent on the mission deployment. The US Navy variant shown in Fig. 8.15 will be crewed by civilian mariners, while the army vessels will be crewed by Army support personnel.

The DoD selected the Austal design, to be built at its new Mobile, Alabama shipyard in 2009. This has been especially built to construct the vessels in a series of modules that are then assembled in a main hall. Construction of the first vessel (JHSV 1 "Spearhead") started in December 2009 and is due for launch in June 2011. The second vessel construction (JHSV 2 "Vigilant") started in September 2010 and will follow a similar construction plan to launch of approximately 18 months. Construction of the following 3 craft is now in advanced planning as the DoD have committed for JHSV 3, and long lead items for JHSV 4 and 5.

In addition to its basic payload capacity the JHSV will be outfitted with hydraulic loading ramps giving it the ability to work at ports with limited access and handling facilities. The aviation facilities will improve support to disaster relief operations or other military special operations. The US Marines used a similar specification vessel built by Austal USA for Ferry service in Hawaii on temporary assignment early in 2010 to deliver relief to Haiti after the earthquake there.

So far we have been considering craft that operate in the Froude number region up to 0.7, essentially displacement operation, with some dynamic lift for craft such as the HSS, but not yet entering the full planing region. The speeds for craft such as HSS and Westpac are high simply because of their large dimensions. We turn now to the full planing mode, with the planing catamaran (PCAT) form.

Planing Catamaran

The design for the PCAT is based upon the planing monohull, see Fig. 6.3c compared to Fig. 4.11. Since the individual hulls can be finer in the fore section, this can improve seaworthiness of the craft, particularly reducing slamming at high speed.

For large displacement or semi-planing catamarans using symmetric hull geometries with an open tunnel the main issue for hull spacing and design of the "tunnel between hulls is wave form interactions". Smaller PCATs, and especially craft used for racing follow the approach from the early Norwegian catamarans where the side hulls are like the longitudinal half of a monohull spread apart and connected with a bridging structure. The space between the "demi" hulls and under the bridging deck forms a tunnel through which air will flow and the velocity can be constrained by adjusting the geometry. Similarly, the hull generated waves below planing speed, and the lower surface deformation above planing speed will also be constrained by the hull tunnel. Both of these effects will generate drag and lift forces. Figure 6.12 shows two body plans of a PCAT, where (a) a type for river operation, with small space between the twin hulls, and shallow air tunnel; (b) shows another type for coastal operation, with wider space and deeper tunnel. We will refer to this configuration as Tunnel Planing Catamarans or TPC for short.

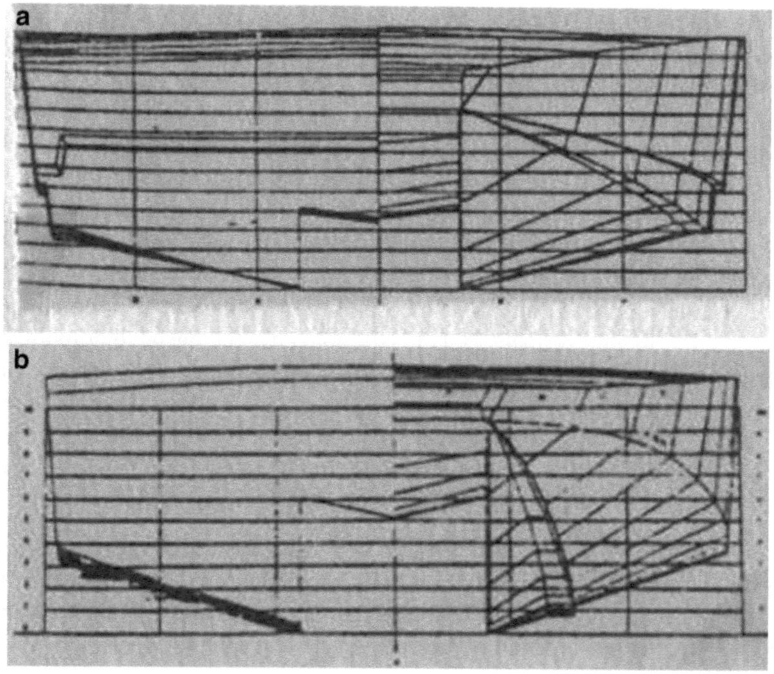

Fig. 6.12 (**a**) River craft. (**b**) Coastal craft

These craft cross sections have a much heavier aerodynamic interaction that semin planing or displacement catamarans due to the different tunnel geometry, and lower internal wave generation when planing due to the vertical tunnel sides. To reduce slamming of the fore body the bridge structure can be formed with a canted vee to part the water surface as the craft pitches down into a wave peak as shown in Fig. 6.12b.

The differences of the tunnel-form planing catamaran compared with monohull planing craft are the following:

- Since the planing surface is split into two slender demi-hulls having lower trim angle, and with a ram air cushion in the tunnel, slamming should be reduced and thus improve seakeeping. The speed loss of such craft in waves should also improve compared with conventional planing monohulls.
- If the tunnel shape contracts from bow to stern along the whole longitudinal section, at planing speed the craft will be supported by the two hull planing surfaces together with support from ram air in the tunnel forming an air cushion to partly support the craft weight, and also generate lubrication effects on the wetted surface at stern helping to reduce friction resistance.
- During the take-off of a TPC, the change of trim angle can be designed to be small as the ram air lift centre is essentially at the centre of area of the tunnel roof, adjusted by the velocity profile generated by the tunnel cross section

changes from bow to stern. The two hulls are much more slender than a monohull so that wave-making is itself reduced at sub-planing speeds. On a conventional planing monohull craft, the centre of lift is sensitive to the change of trim angle and planing surface wetted length which diminishes and moves significantly aft as speed increases. Wave-making is more significant sub-planing on a monohull, so that craft trim is larger, and the drag hump is also more severe.
- For the river craft example in Fig. 6.12a the hull-like forward tunnel roof may provide additional lifting support just below planing speed and assist to depress the resistance hump aiding takeoff into planing.
- Higher transverse static and dynamic stability than an equivalent monohull.
- The slender planing surfaces of TPC will not cause the unstable dolphin motion typical of monohull craft that are not optimally trimmed. In addition widened separation between the two hulls increases the distance between two water propellers, thus, improving propulsive efficiency and manoeuvrability.
- Increased deck area for accommodation similar to the displacement type fast catamaran.

The PCAT uses dynamic lift to support craft weight so slamming loads will still be high in case of rough sea operation at high speed, compared with the displacement type high-speed catamaran. Like the high-speed planing monohull, it can only be applied practically for small high-speed craft, for example:

- Small passenger ferry craft
- Recreation craft and racing boats
- Border patrol craft
- Fast interdiction craft against pirates, smugglers, and illegal fishing

The leading particulars of some TPC, made by Cougar Co, UK, are listed in Table 6.3. Two examples of high-speed PCATs are shown in Fig. 6.13a, b below.

Table 6.3 Leading particulars of example fast catamarans from Cougar Marine

Type	CAT900	CAT1400	Cougar20	CAT2000	CAT2100
Loa, m	9.70	14.30	17.68	19.88	21.60
Lwl, m	9.20	14.00	17.68	19.80	21.0
Boa, m	2.89	5.00	6.71	6.24	5.50
Hull depth, m	1.27	1.94	2.42	2.84	2.60
Draught, m	0.78	1.20	1.37	1.38	1.5
Power, kW	312	735	1,863	2,881	4,413
Speed, knots	42	38	47.75	50	51
Displacement, t	4.80	12.40	20	36	53.23
Frl	2.27	1.67	1.87	1.85	1.83

Fig. 6.13 (a) JBS racing cat. (b) Cougar "Kaama" racing cat

Racing catamaran craft have been able to reach astounding speeds even in open coastal waters. The boat shown in Fig. 6.13a, has taken the offshore speed record at over 300 kph. This is as fast as water speed records in flat calm conditions in the middle of the twentieth century with the three-point hydroplane (see below), and indicate the major achievements of designers with hydrodynamics and aerodynamics in the beginning of the twenty-first century. To operate in a stable manner at these speeds the hull above-water form needs to be carefully designed for stable aerodynamics, as much of the craft weight is supported on the ram air cushion from the tunnel. Propulsion of this particular craft is gas turbines driving surface riding propellers. The planing surfaces of the hulls include several steps as well as spray rails on the deep vee lower surfaces. The combination of all this allows the craft to keep a stable trim while travelling rapidly over what has become a "rough" surface in normal conditions for racing.

Wave Piercing Catamarans

In contrast to the PCAT form we have addressed above, the WPS is a development of the slender water plane displacement form. In this case, the bow of the two side hulls is formed into a submerged finger-like shape so that when running in waves the craft has lower tendency for the bow to be lifted by the oncoming wave peaks. Instead, the craft slices through the waves and has a much smaller pitching motion. This in turn reduces induced wave-making and the speed loss in a seaway. Since the bow slices through waves, slamming forces are also reduced. Since the catamaran hulls are very slender, designers of WPC form the bow of the craft centre body into a bow form also. When in heavy seas, this bow structure provides buoyancy to limit the immersion of the catamaran hull bows.

The hull geometry of wave piercing craft is a hybrid of CAT, small water plane area twin hull craft (SWATH), and Deep V monohull high-speed craft. The advantages and disadvantages of the three craft can be summarized briefly as follows:

Advantages	Disadvantages
Deep Vee Monohull craft	
• Fine seaworthiness	Lower transverse stability in case of high slenderness
• Good high-speed performance	High motions and accelerations in a seaway
	Low volume for payload
HSCAT	
• Simple structure	Challenging structural design for large craft
• Economic to build and maintain	Poor seaworthiness in beam seas
• Large usable deckhouse area	
• Medium speed	
SWATH	
• Marvellous Seaworthiness	Large wetted area, friction drag at high speed
• Large deck area	Poor longitudinal stability
	Deep draught
	Sensitive to the weight distribution and changes
	Complicated power transmission

The goal for the WPC configuration is to try to use the positive attributes of the three craft types summarized above, and avoid the disadvantages, so as to create a craft with a better overall performance. There has to be a trade-off for several of the attributes since in order to achieve a hull form that gives better high-speed performance in a seaway the geometry has to become more complex. The structural design

Fig. 6.14 Front view

to resist racking forces in the central structure from the more widely spaced hulls than other catamarans has tested designers innovation to produce a stiff structure that is also light and still allows large enough open spaces for vehicle roll on/roll off loading and passenger saloons that are comfortable.

The key challenge early wave piercing craft faced was that because they were able to maintain speed in higher sea states they tended to be driven harder as the crew felt comfortable; this meant that the main hull structure was being worked harder, and so fatigue cracking appeared on some craft which had to be corrected with additional strengthening. Once this was fed back into later craft designs, the problem was able to be corrected in build.

Over the last 2 decades increasing optimization using finite element analysis for structural design combined with feedback from craft in service has enabled refinement by the builders so as to maintain economy in the build and reliability in operation.

The following four figures show examples of WPC craft. Figure 6.14 shows the front view of a WPC "Condor Vitesse" in service, Fig. 6.15 Shows "Condor Vitesse" at speed. Figure 6.16 shows an earlier WPC running in waves, and one can see the craft bow pierces through the wave crests, and Fig. 6.17 shows a typical WPC body plan.

Main features of the WPS configuration are:

Slenderness: the catamaran is slender and with hull slenderness ratio $L / \nabla^{1/3} = 9 - 11$, and $L/b = 10–19$, $B/2b = 1.2–2.3$ (where L, b, Δ, represents length, width, and displacement respectively of a side hull, and B the catamaran beam also see Table 6.4). It is a rather slender and deep draught compared with conventional catamaran; therefore, the entrance angle at waterline is small for running in wave piercing manner.

Fig. 6.15 Condor 10

Fig. 6.16 WPC craft at speed

Low freeboard and thin struts: The freeboard of WPC is low, particularly at the bow, and the bow reserve buoyancy is reduced, so decreasing wave perturbation as well as heaving and pitching motion in waves. The hull configuration above the design waterline is rather different from that on ordinary high-speed catamaran, with thinner configuration, and smooth transition into the strut. A central bow structure suspended above the normal waterline including wave interaction is designed to provide buoyancy in the case of severe pitching motions.

Shape of bow/stern parts: A deep vee cross section is usually used for the bow (Fig. 6.17). The stern shape of a WPC is similar to that on an ordinary high-speed

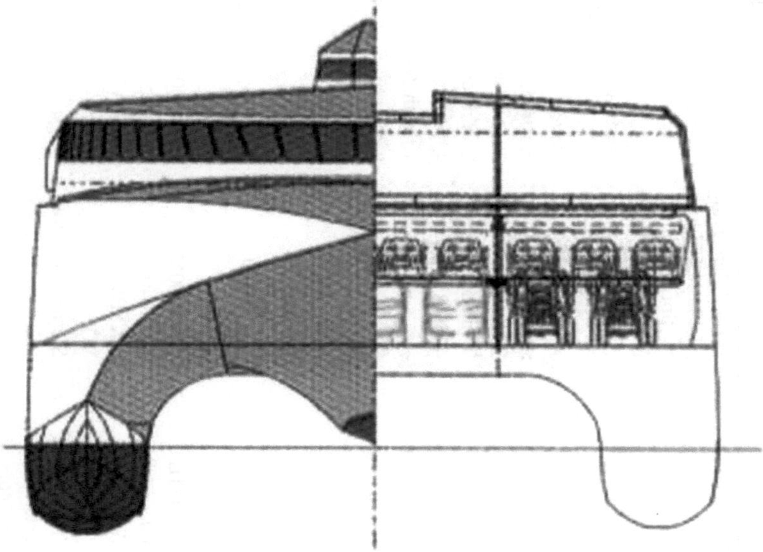

Fig. 6.17 WPC transverse section

catamaran. Since WPC speed is higher with water jet propulsion installed, the stern is of transom type with small dead rise angle. The keel line is inclined from stem to stern and drops below the craft baseline at the stern to increase the cross-sectional area under water and increase the damping force in longitudinal motion to prevent the bows emerging from the water surface in waves. At the same time, the half angle of entrance of the vee form lower surface is reduced ($\alpha / 2 \leq 6^0 - 10$) compared with a PCAT, and wave resistance is reduced, see Fig. 6.18 for a detail view of the WPC bow slicing through waves.

Large hull separation: The craft beam to hull beam ratio, B/b, where B is WPC beam, and b is the beam of the side hulls, is as high as 5.5–6 instead of 3–4 for high-speed catamaran; thus, the interference effect between the hulls is small or even favourable. Transverse stability will be improved even in case of higher superstructure, and also craft roll angle will decrease in waves. In addition, the arrangement of passenger cabin can be improved. The connecting structure between the hulls will have rather higher stresses due to out of balance heaving and pitching perturbation forces (twisting and bending) from the hulls and this has to be taken into account in the craft design.

Connecting bridge & central hull: In general, the transverse section of connecting bridge is of arch type (Fig. 6.19), to reduce the wave impacting loads, give higher transverse stiffness and minimize wave induced vibration of the total structure in waves and associated fatigue loading.

Simple power transmission: WPC hull beam is wide at the stern, and in addition, the hull depth is sufficiently high to locate the main engines into the hulls, so as to use direct power transmission to water jet propulsors enhancing the transmission efficiency and simplifying the outfit and its maintenance.

Table 6.4 Leading particulars of example current large catamaran ferries

Name	Sawqrah, Hallaniyat, Masirah	MGC66	Design proposal	Jean de Valette	Paria Bullet, Trini Flash, Calypso Sprinter, Carnaval Runner	Silver express
Design	WP cat 52S, wavepiercer	Incat 112, wavepiercer	Incat 85, wavepiercer	Auto express 107	Austal 41 m	Austal 47 m
Builder	Rodriquez	Incat Tasmania	Incat Tasmania	Austal	Austal	Austal
Country	Italy	Australia	Australia	Australia	Australia	Australia
Year	2010	2009	Design	2010	2010	2011
Operation	Oman	Japan	tba	Malta	Trinidad	Caribbean
Passengers	116+10	1,182+18	630	920+24	405+8	437+7
Cars	22 (or 12+3 trucks)	417 (or 195+32 trucks)	181	156 (or 40+28 trucks)	–	10 with 364 pax
Displacement, t	1,228	10,800	n/a	8,045	562	n/a
Loa, m	52	112.6	84.6	106.5	41.2	47.0
Boa, m	15.5	30.5	26.6	23.8	10.9	11.1
Hull depth, m	5.2	7.9	n/a	9.4	4.3	4.0
Draft, m	2.5	3.9	3.4	4.9	2.0	1.82
Speed, knots	40	35 (39 max)	36 (40 max)	39	37	32
Main engines	4×MTU 16V4000M71	4×MAN 28/33D20V	4×MAN 28/33D12V	4×MTU 20V8000M71L	4×MTU 16V2000M72	2×MTU 16V4000M71
Power, kW	4×2,465	4×9,000	4×5,400	4×9,100	4×1,440	2×2,465
Propeller	4 KMW71SII WJ	4 LJX1500SR1 WJ	4 Wartsila 11000SR WJ	4 KMW125SIII WJ	4 KMW56AIII WJ	2 KMW71SIII WJ
Stability fins	T foil under central bow	T foil under each bow, trim tabs stern	Bow T foils, and stern intercepters	2 bow T foils and stern interceptors	Stern interceptors	–
Structure	Al alloy	Al alloy	Al alloy	Al alloy	Al alloy	Al alloy
Frl	0.91	0.60	0.71	0.62	0.95	0.77

Name	Red Jet 1 Red Jet 2	Zao Quing	Ryadh and Cairo	Auto express 113	New Ferry LXXXV11, LXXXV111	Westpac	JHSV
Country	UK	Australia	Australia	Australia	Australia	Australia	USA
Builder	FBM Marine	Austal	Austal	Austal	Austal Tasmania	Austal	Austal USA
Design	FBM	Austal 42	Express 88	Express 113	Austal 48	Express 101	JHSV
Operation	UK	Hong Kong	Egypt	Sweden Bornholm	Hong Kong	USA	USA
Year	1989	1997	2008	2011	2008	2005	2009>
Passenger+crew (deck level)	120 main	352+12	1,200+18	1,400+35	418+8 crew	900+26	452+41
Cars			120+15 trucks	357		152 hummers	Flexible mission decks
Loa, m/LWL	31.5	42.1/37	88.0/77.3	112.6/101.3	47.5	101/86.2	103
Boa, m	8.4	11.5	24	26.2	11.8	26.7	28.5
Deck area, m^2	n/a	n/a	n/a	n/a	n/a	2,490	1,800
Hull depth, m	2.5	3.7	8.25	8.5	3.8	9.4	n/a
Draft, m	1.1	1.3	4.3	4.85	1.6	4.27	3.83
Deadweight, t	n/a	n/a	555	1,000	n/a	790	1,858
Speed light, kt	35.5	35	40	40	43	36	43
Speed loaded, kt	32.5	34	37.7	37	40	33	35
Engines	2 MTU 12V396 TE84	2 MTU 16V396 TE74L	4 MTU 20V8000 M71R	4 MAN 20V28/33D	4 MTU 16V4000M70	4 Caterpillar 3618	4 MTU20V 8000M71L
Power, kW	2×1,360	2×1,980	4×7,200	4×9,100	4×2,320	4×7,200	4×9,100
Propulsion	2×MJP 650, Waterjets	2 KMW 71SII, Waterjets	4 KMW 112SII, Waterjets	4 KMW 125SIIINP, Waterjets	4 KMW 63 SII, Waterjets	4 KMW 125SII, Waterjets	4×Wartsila WLD1400SR, Waterjets
Stabilizers	–	–	Bow T foils	Bow T foils	–	n/a	Bow side foils Stern interceptors
Frl	1.04	0.89	0.70	0.62	1.02	0.54 (0.59)	0.57 (0.70)

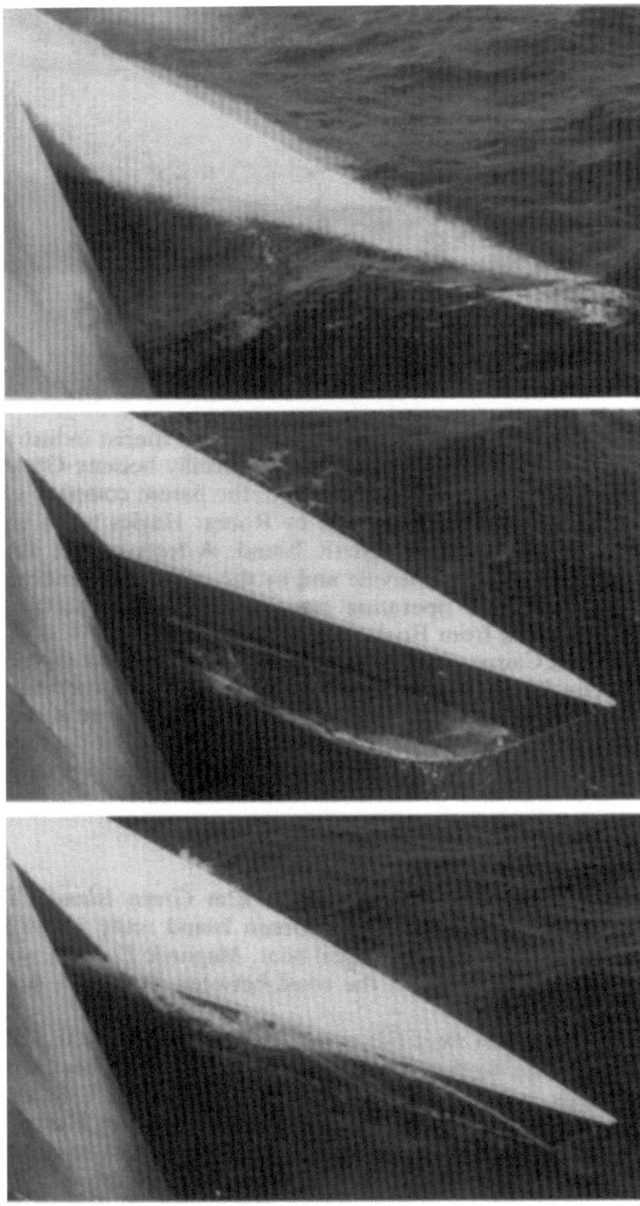

Fig. 6.18 Bow slicing waves

Heavier hull structure The disadvantage of the WPC configuration is that it has a more complex hull structure at higher cost than the other catamaran configurations introduced above. Due to the lower reserve displacement above waterline forward of amidships, they need to be fitted with automated stabilizer fins to maintain seaworthiness, so also increasing the cost.

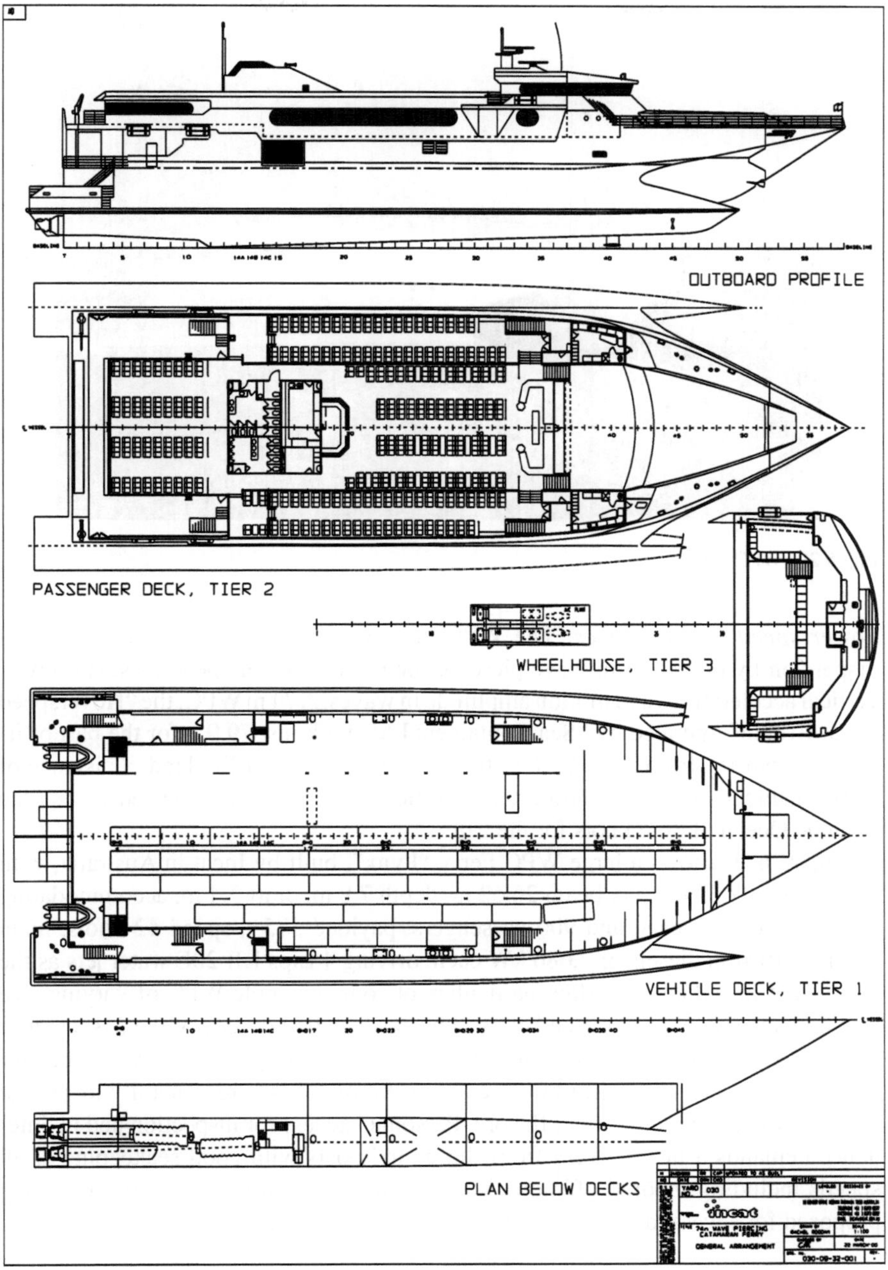

Fig. 6.19 WPC Condor 10 general arrangement

The key advantages of WPC configuration are:

High speed: the craft will be operated at higher speed than a traditional catamaran, say, Frl = 0.75–1.1, and higher, normally up to 35 knots and more.

Fig. 6.20 WPC Lynx

Lower speed loss in waves: The WPC has lower added resistance from wave dispersion by operating in wave piercing mode, with lower speed loss, and lower vertical accelerations and motion amplitude in waves. A 74 m WPC, the "Hoverspeed Great Britain" operated by "Sea Containers Ltd" took just 79.9 h for the part of its delivery journey from New York to the south-west coast of England, a distance of 5,400 nautical miles, and broke the historic record of average speed across the Atlantic Ocean of 36.6 knots for the trip.

Figure 6.20 shows a large WPC Ferry "Lynx", built by Incat in Australia, with max length 97.22 m, max beam 26.62 m, depth 7.8 m, draft 3.4 m, accommodating 380 buses or 267 cars, and 900 passengers, payload 375 t, speed 42 knots, main engine 4 Ruston diesels of 7,080 kW each, driving 4 Lips LJI 20E water jets as the propulsion installation Leading particulars of some example WPC of varying size are shown in Table 6.5 at the end of this chapter. The first decade of the 21st century has seen a new utility market open up for catamaran craft, generated by the need for access to offshore windmill power generator farms. These are generally located at shallow water sites some way offshore. Personnel access for inspection and maintenance demands a fast delivery from shore, and boats with good seakeeping at all speeds including moored offshore. Variants of wavepiercing geometry are being proposed for several designs, as listed in Table 6.5 below.

Trimarans

The starting point for fast trimarans might be considered as the three point hydroplanes that were built in the 1930s for attempts on the water speed record [1-2]. Sir Malcolm Campbell's first such craft was the Bluebird K3/K4, built in 1939

Table 6.5 Examples of wave piercing catamarans (WPS)

Vessel name	Condor 10	Stena Lynx III	Condor express	Bencomo express	LD Lines Norman Arrow	Hull 069 (LNG/diesel)	Fei Ying
Type	Incat 74 m	Incat 81 m	Incat 86 m	Incat 96 m	Incat 112 m	Incat 99 m	AMD 150
Builder	Incat Tasmania	Incat Tasmania	Incat Tasmania	Incat Tasmania	Incat Tasmania	Incat Tasmania	Hang Tong
Country	Australia	Australia	Australia	Australia	Australia	Australia	China
Completed year	1993	1996	1996	2000	2009	2012	1998
Country operated	UK	Ireland	UK	Canary Islands	UK	Australia	China
Passenger	450	700	900	920+21	1,180+20	1,000+	180
Car	84	173	200	245	417 or 195 and 567 lane m truck space	153	–
Deadweight, t	230	320	415	708	1,000	450	21
Loa, m	73.6	81.1	86.6	95.5	112.6	99.0	25.0
Boa, m	26.0	26.0	26.0	26.6	30.5	26.6	9.57
Sidehull breadth, m	4.33	4.33	4.33	4.01	5.8	4.0	n/a
Draft, m	3.1	3.0	3.5	4.0	3.93	3.4	1.8
Max speed, knots	43	44.0	48	46.7	45.0	55.0	31
Service speed, knots	37	38.0	40	38	35–38	50.0	28
Main engines	4 Ruston 16RK270	4 Ruston 16RK270	4 Ruston 20RK270	4×Caterpillar 18V3618	4×MAN 28/33D	2×GE LM2500 gas turbine	2×MWM TBD616V12
Power, kW	4×4,050	4×7,080	4×7,080	4×7,200	4×9,000	2×23,340	2×839
Propulsion	4 WJ Lips LJ115DX	4 WJ Lips LJ135DX	4 WJ Lips LJ145D	4×WJ Lips150/3D	4×WJ Wartsila LJX1500SR	2×WJ Wartsila LJX1720	2×FPP
Ride control	Maritime dynamics 2 bow T foils and stern fins	Maritime dynamics 2 bow T foils and stern fins	Maritime dynamics 2 bow T foils and stern fins	Maritime dynamics 2 bow T foils and stern fins	Maritime dynamics 2 bow T foils and stern fins	Maritime dynamics 2 bow T foils and stern fins	n/a
Structure	Al alloy	Al alloy	Al alloy	Al alloy	Al alloy	Al alloy	Al alloy
Frl max	0.82	0.803	0.85	0.79	0.70	0.90	0.99

Fig. 6.21 Bluebird K4 at speed

which could run at up to 145 mph (125 knots). This was a craft with two planing surfaces forward and a single surface aft from which propellers were mounted. Initially powered by a reciprocating aero engine, the craft was rebuilt just after the Second World War with a De Havilland Goblin jet for power. Initially unstable at high speeds, the craft was modified to run as a prop rider at the stern, Fig. 6.21, and was able to achieve a speed as high as 170 mph (147 knots) on calm water, before hitting a submerged log and being wrecked.

The replacement craft, also had two sponsons with planing surfaces forward (referred to as skis for these record breakers) and a single support point at the stern of the central hull, this time being powered by a Beryl jet engine, Fig. 6.22. The new Bluebird, K7, driven by Donald Campbell was able to achieve 200 mph (174 knots) in 1955. In 1964, refitted with a higher power Orpheus jet engine she raised the record to 276 mph (240 knots) in Australia. In 1967 a new speed record attempt was made on Coniston water in the UK Lake District, and unfortunately the craft flew up and flipped backwards into the water while travelling close to 300 mph (260 knots), wrecking the craft and killing Donald Campbell. The story of this development is chronicled in [6-5].

At this speed, the greatest problem is to keep the hull on the water while also dynamically stable. We have introduced the WIG in Chap. 3. Most of these craft fly at low speeds compared to the record breaking tri-point hydroplanes referred to above, yet are flying above the sea surface. The speed record hydroplane development in the middle years of the twentieth century illustrates the high end extreme

Fig. 6.22 Bluebird K7 record in Australia

Table 6.6 Leading particulars of craft and designs for offshore wind farm work

Name	Fintry 23	Wind express	TSV 24	Offshore unlimited	FOB swath 1
Design	Carbocat	Austal	BMT Nigel Gee	Incat Crowther	Ola Lilloe Olsen
Builder	Kockums	Austal	tba	RDM	Måloy Verft
Country	Sweden	Australia	UK	Australia	Norway
Year	2010	Design	Design	2010	2010
Operation	Baltic	tba	tba	Bass strait	N sea
Application	Wind farms	Wind farms	Wind farms	Offshore supply	Wind farms
Passengers	12–24	52+4	21+3	38 (+14 berths)	36
Cars	–	20 t freight	n/a	Freight n/a	10 t freight
Displacement tonnes	n/a	n/a	n/a	n/a	125 as cat 149 as swath
Loa, m	23.6	28.5	24	28.7	26.6
Boa, m	8.7	9	7.7	8.5	9.6
Hull depth, m	3.4	na	na	3.45	4.4
Draft, m	1.2	1.9	1.0	1.2	1.75, 2.75
Speed, knots	25 (31 max)	26	26	25 (30.5 max)	24, 14.5
Main engines	2×MTU 8V2000M72	2×MTU 10V2000M72	2×Caterpillar C18ALERT	2×Caterpillar C32ALERT	2×MAN D2482LE410
Power, kW	2×720	2×900	2×533	2×1,080	2×749
Propeller	2×WJ	2×FPP	2×FPP or WJ	2×FPP	2×CPP
Stability fins	–	n/a	n/a	–	2 fwd
Structure	CFRP	Al alloy	Al alloy	Al alloy	FRP
Frl	0.84	0.80	0.87	0.77	0.76 (0.46)

for HPMV. Since that time the hydroplane has continued to develop particularly for short circuit racing powered by outboard motors. They are able to maintain stable operation over the wake of others and performing sharp turns operating in the speed range of 60–80 knots, see resources at the back of the book for web sites about hydroplanes.

Given sufficient power, and a short enough mission it is possible to design a form that is stable at high speed, at least on calm water, or limited chop. The challenge in recent years is whether this form can be put to use with a significant payload for commercial purposes, and whether it might have any advantages over the catamaran forms reviewed above.

In fact the trimaran form has been developed along different lines for military and commercial HPMV use. The principle is the same—provision of stability to a slender central hull. The racing craft central "hull" became diminished to a small ski, or even the surface riding propeller itself for the record breaking craft at Froude numbers in the range 8–10 based on hull length. For craft operating in the Froude number range of 0.7–1.1, the hull may be semi-planing but will still remain substantially in contact with the water surface. When we consider a fast commercial monohull, or a military vessel such as a corvette or a frigate, if we wish to improve performance what can we do while retaining the basic central hull?

The answer was to place "sponsons" either side of the hull, and allow the hull to be more slender—this is where we came in with the stability problem for very fine planing monohulls of course. The first significant project to try this out was commissioned by the British Royal Navy, a trimaran to demonstrate the effectiveness of the concept to possibly replace the older monohull Royal Navy Type 23 frigates as part of the "Future Surface Combatant" Project. The issues with existing monohull designs that the project was to investigate were as follows:

- How to reduce wave-making compared to existing ships at high speed (Frl > 0.5), due to relatively wide beam aft, and relatively low length/beam ratio adopted for stability
- A search for greater space both on deck for helicopters and holds below deck for modern weapons outfit
- Improved motions at speed in a seaway to allow helicopter operations in heavier weather, and to improve operability of the weapons outfit

The approach chosen to improve on these limitations was by extending the main hull in length, to give a high length/beam to reduce wave-making resistance; and to add two sponsons at each side of the main hull amidships to improve the transverse stability, as shown in Fig. 6.23, RV Triton.

RV Triton was launched in 2000 and put through a series of trials in the period up to 2005, including helicopter operations, berthing, and offshore replenishment operations with monohull support vessels. The vessel is 97 m length, 22 m overall beam, 800 tonnes displacement and maximum speed 20 knots, cruise speed 12 knots, and with a range of 3,000 nautical miles. The trimaran was not eventually chosen for the Royal Navy development programme but did give valuable knowledge for operation and design of these vessel types. RV Triton was actually bought by the Gardline Marine group in 2005 and since end 2006 it has been operating under lease by the Australian Customs service as an offshore interdiction vessel.

Fig. 6.23 RV Triton

In the period Triton was being developed, the fast ferry builder Austal also investigated Trimaran configurations for commercial ferry operation, targeted at improved seaway performance compared with the company's main product, the fast catamaran. Austal found that performance was improved by having the sponsons at the stern of the vessel rather than amidships. Figure 6.24 shows the military version of Austal trimaran design recently adopted by the US Navy for comparison with Triton. The Austal Trimaran is designed for a service speed of 36 knots, significantly higher than Triton. The main features of this kind of trimaran are as follows:

- Low wave-making resistance at high speed, due to the slender main hull, and favourable interference of wave-making between the main hull and two side sponsons.
- Adequate transverse stability due to the sponsons without being too stiff, so giving good response in oblique seas.
- Large deck space, particularly aft, which is convenient for helicopter operations. Triton's configuration with the sponsons amidships gave less space in this respect.
- Improved the seaworthiness with small roll angle due to high damping of the roll from the sponsons, as well as smaller pitch motion and speed loss due to the slender main hull.

Austal's first customer for the trimaran was Fred Olsen Ferries in the Canary Islands. The Benchijigua express is 127 m, length, 30.4 m beam, with a displacement of 2,600 tonnes. Delivered in 2005 the craft can carry up to 341 cars as well as 1,291 passengers on its route between Tenerife, and the islands of La Palma, La Gomera and El Hierro in the Canary islands, Fig. 6.25.

Fig. 6.24 Austal Trimaran

Fig. 6.25 Fred Olsen Trimaran Ferry Benchijigua express

In order to obtain favourable wave-making interference between the main hull and sponsons, and an optimized general arrangement for the ship, the two possible arrangements for the sponsons were investigated by Austal before settling on the aft

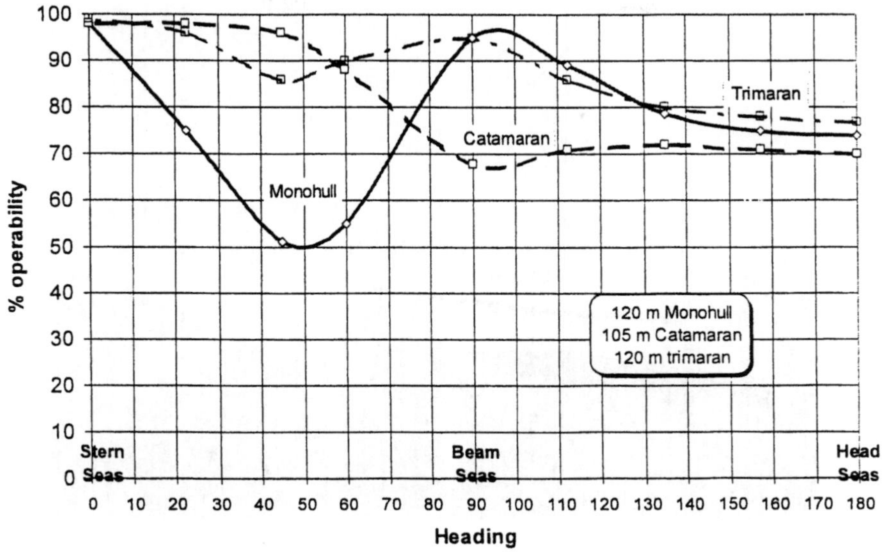

Fig. 6.26 Operability analysis for Western Pacific

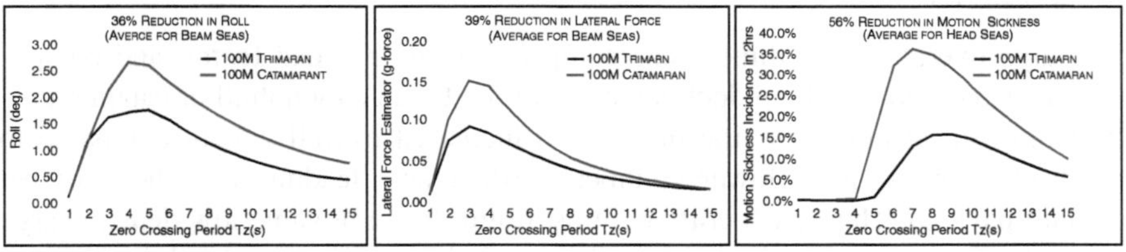

Fig. 6.27 Motion data comparison

configuration, either arranging the sponsons aft or in the middle similar to HMS Triton. The aft configuration was much more convenient for the machinery installation where waterjets would be installed in all three hulls, and the hydrodynamic interaction was found to be more favourable for the higher Froude number operation of the high-speed ferry. Once the stern configuration was selected, a series of model tests was carried out at the Hydrodynamics Research laboratory MARIN in Holland to optimize the geometry.

Figure 6.26 shows the result of an operability analysis of the equivalent monohull, catamaran and trimaran (stabilized monohull) designed for 1,000 t payload, in the Western Pacific area, for a selected motion criteria. The superior operability of the sponson stabilized trimaran is clearly illustrated. The response of a catamaran and trimaran at different headings relative to waves is shown in Fig. 6.27, demonstrating the reduction in roll and lateral force in beam seas and acceleration in head seas with associated reduction in motion sickness.

Fig. 6.28 Austal 102 m Trimaran Ferry

So far so good, but what are the limitations? The disadvantages of the trimaran are as follows:

- Large overall vessel dimensions compared to the payload factor, and complicated geometry, making docking more difficult than a monohull or catamaran.
- Complicated hull structural design and increased overall weight compared to deadweight payload (lower volumetric efficiency), leading to higher cost per unit payload. Thus, such craft are most useful for high value low density cargo such as passengers, and tactical military missions rather than hardware (cargo) projection to a target.

In an open sea environment such as the Canary Islands, the operability was an important factor, and led to its selection over a catamaran. The operability was also a key factor for the US Navy, which became interested in the vessel for its Littoral Combat Ship (LCS) programme aimed at up to 55 new high-speed vessels. In 2005 Austal, working with General Dynamics was given a contract to build a prototype as part of a competitive evaluation for LCS concept. LCS2, USS Independence was built at Austal's Mobil shipyard in the USA, and put on a trials programme from 2009.

The trimaran concept was selected later in 2009 by the US Navy and a contract let to the General Dynamics consortium for the second vessel USS Coronado, to be built at General Dynamics Bath Iron Works shipyard. The US Navy will continue building the trimaran LCS class with a ten ship programme that was committed to at the end of 2010. We discuss the programme further in Chap. 8. Austal have continued development of the Trimaran, and have built a second 102 m class ferry,[6-6], Fig. 6.28, and have prepared a design for a larger military craft, the MRV 126 shown in Fig. 6.29. Table 6.7 gives data for fast trimarans as well the ASV Craft and M Craft discussed in Chap. 7, for comparison.

Fig. 6.29 Austal 126 m Trimaran

Pentamaran

This design was initiated by a British research and engineering company BMT Defence Service Ltd, with the aim at improving endurance in high sea states, and superior seakeeping quality.

The main hull is optimized for drag at a given deadweight and speed, while the four sponsons—two aft and two forward—serve to eliminate parabolic rolling while minimizing drag. The two aft sponsons are sized to maintain an adequate yet shallow immersion below the water line. The sponsons mounted forward are designed to sit just clear of the water, becoming immersed only during a rolling motion (Fig. 6.30).

The pentamaran in effect has a leg in each corner, and as the vessel heels, one sponson emerges, so another immerses to restore stability. This reduces resistance compared with the trimaran, together with improved seakeeping, at the expense of rather more complex sponson design. The vessel itself is not sensitive to weight, but the key thing for the pentamaran is to operate at a constant draft and narrow range of trim to ensure the correct immersion of the sponsons, and this means that the ships must take on water ballast to compensate for variable load. This creates a penalty in economic performance compared to the trimaran, and may mean that the concept is only interesting where even higher speed and performance than the trimaran is required. This may be a narrow window of application before one enters the planing regime and different drivers to optimisation. Figure 6.31 shows general arrangement of the Pentamaran Frigate by BMT, to meet future requirements for speed flexibility and "Swing" role capability.

Table 6.7 Leading particulars of SSTH craft

Vessel names	Ocean arrow	Stena Carisma	Stena explorer Stena Voyager HSS discovery	Seafighter Fast sea frame FSF-1	USS sea shadow
Ship type	70 m SSTH	Stena, HSS 900	Stena, HSS 1500	X craft	IX-529
Country	Japan	Sweden	Sweden	USA	USA
Completed year	1998	1997	1996, 1997	2005	1985
Builder	IHI AMTEC	Westamarin	Finnyards, Rauma	Nichols Bros, Washington	Lockheed Martin
Passenger capacity	430	900	1,520	–	–
Crew	3	n/a	n/a	50	4+8
Cars	51	210	360	Freight 450 t	–
Length overall, m	72	89.75	126.6	79	50
Breadth, m	12.9	30.47	40.0	22	21
Hull depth, m	3.9	21.0	27.5	n/a	n/a
Draft, m	1.6	3.9	4.8	3.6	4.6
Max speed, knots	30+	45	51	50+	28
Speed loaded	30	40	40	40+ (SS4)	13
Main engines	2 MTU 16V595TE70L	2×ABB-STAL GT35 Total 33.5 MW	2×GE LM2500 2×GE LM1600 Total 68 MW	2 GE LM2500 2 MTU16V 595TE90	Diesel-electric
Power, kW	2×3,950	2×16,750	2×23,380 2×11,000	2×23,380 2×4,350	n/a
Propulsion system	Propeller	2×KMW WJ	4×KMW WJ	4×WJ KMW125SII	2×CPP
Hull material	Al alloy	Al alloy	Al alloy	Al alloy	Al alloy
Navigation region	Inter-island Japan, rough water	Kattegat	North Sea, Irish Sea, Brazil	Global Range 4,000 nm	SS4 op SS6 max
Frl (service)	0.58	0.69 (0.78)	0.58 (0.75)	0.73 (0.92)	0.3 (0.65)

Fig. 6.30 BMT Pentamaran Frigate proposal

Super Slender High-Speed Catamaran

The design concept of the super slender high speed catamaran is similar to the trimaran and pentamaran, as mentioned above though in this case simply extending the side hulls to greater slenderness, aiming at improving both resistance characteristics and seakeeping quality. This was a development by IHI AMTEC corporation, and Tokyo University in at the beginning of 1990s. Figure 6.32 shows the profile of their super slender high-speed twin hull (SSTH) passenger ferry Ocean Arrow, and the leading particulars are shown in Table 6.8 on page 245. At 72 m length and 30 knots service speed this craft operated in the sub-planing regime. The design was aimed at minimizing installed power for economy, while giving a smooth ride for its passengers. The ferry operated a route between Kumamoto and Shimabara in Japan on a 30 min schedule, halving the time taken by the ferry it replaced in 1998.

Small Water Plane Area Twin Hull Craft

We have looked at a number of configurations of multihulled vessels aimed at high-speed service with minimized costs while giving a comfortable ride for the payload of passengers. Surface supported craft; even where the wave piercing concept is employed clearly has limitations for comfort in a seaway. The SWATH is a search

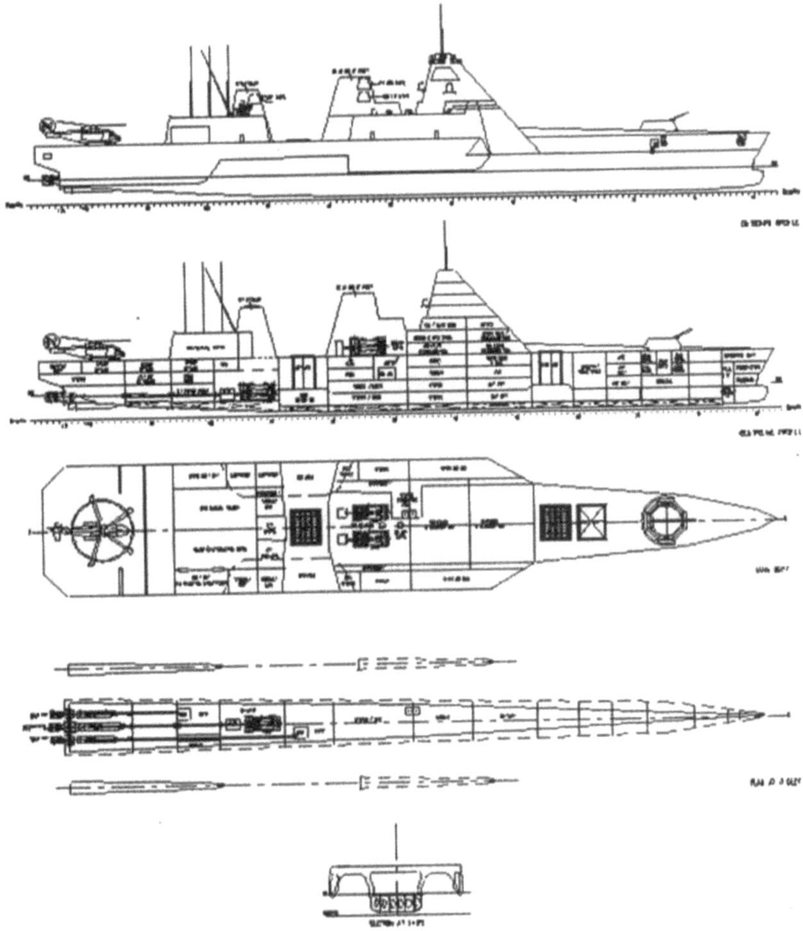

Fig. 6.31 GA of Pentamaran

Fig. 6.32 Super Slender Twin Hull Ferry "Ocean Arrow"

Table 6.8 Leading particulars of Trimarans and others

Name	HMS triton	Bechijiqua express	LCS-2 independence	Austal 102	ASV M65 IPS	M80 Stilletto
Country	UK	Australia	USA	Australia	Norway	USA
Year	1997?	2005	2008	2009	2010	2006
Builder	VT group	Austal	Austal	Austal	Effect Ships Int.	M craft
Vessel type	Trimaran	Trimaran	Trimaran	Trimaran	Air support	Multihull
Application	Warship	Ferry	Warship	Ferry	Demonstrator	Military
Crew/passengers	28 + 14	35 + 1,291	40 + 36	18 + 740	2 + Tba	3 + 12
Cars	–	341	–	245 or mix	–	–
Displacement, t	n/a	2,600 (1,000 t dwt)	2,784 (608 t dwt)	n/a	19–22.5	na
Loa, m	97	126.65	127.4	102	20	27
Boa, m	22.45	30.4	31.4	27.4	5.4	12
Depth, m	n/a	8.2	7.5	7.6	na	5.6
Draft, m	3.0	4.0	3.96–4.5	4.5	1.5	0.8 (0.46)
Speed light, kt	20	42	50	39	33	55
Speed loaded, kt	12	36	44	35	33	35 SS4
Engines	Diesel electric	4 MTU20V 8000	2 MTU20V 8000 2 GE LM2500	3 MTU20V 8000M71	2 × Volvo Penta IPS600	4 CAT c32
Power, kW	n/a	4 × 8,200	2 × 9,100 2 × 23,380	3 × 9,100	2 × 320	4 × 1,213
Cushion lift power system					1 × Volvo Penta D3 at 80 kW for air cushion lift	
Propulsion	Single shaft propeller	2 × KMW125SII 1 × KMW180BII	4 WJ LJX1300 Wartsila	3 WJ LJX1300 Wartsila	2 × FPP	4 WJ
Structure	Steel	Al alloy	Al alloy	Al alloy	FRP	FRP
Frl	0.21	0.61	0.73	0.63	1.21	1.08–1.74

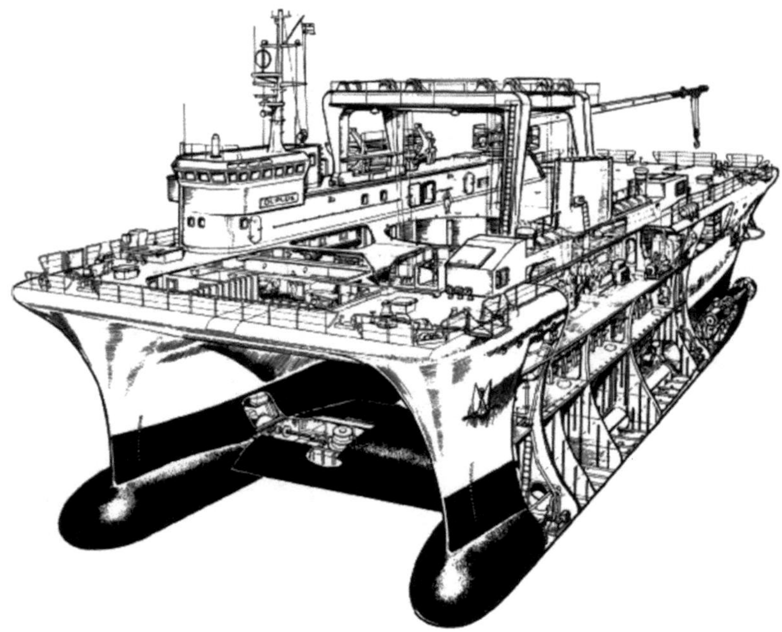

Fig. 6.33 Duplus SWATH

in a different direction for high performance [6-7,6-8]. The concept of SWATH can be traced from the submarine. Once a submarine submerges to a depth greater than one wave height of the surface waves, wave-making of the vessel, and also wave interference from the outside to the vessel are reduced almost to zero, thus improving its seakeeping performance. If the buoyancy for a ship could be completely submerged like a submarine and connected to a support platform above the water with vertical struts, there may be improvements to be gained, in seakeeping and speed in a seaway (Table 6.8).

This basic idea was originally patented by C.G. Lundberg in 1880. He proposed a platform supported from a single submerged buoyant cylindrical body. In 1946 Frederick Creed was granted a patent for a platform supported from two parallel submerged cylindrical bodies, aimed at use as an aircraft carrier. While offered to the British Admiralty, it was not accepted. In 1967, Dr Reuven Leopold of Litton Industries patented a ship hull supported clear of the water by two parallel submerged cylinders and vertical struts. Finally, in 1968 J.J. Stenger and the Boele Shipyard in Holland designed and built the offshore support vessel Duplus using two cylindrical hulls supporting a rectangular above water hull to support offshore oil developments (Fig. 6.33). The Duplus was not an HPMV at 8 knots, but did demonstrate the principle successfully, and encouraged developments by several other shipyards. In Japan Mitsui began work on the semi submerged catamaran which resulted in the 26.5 knot, 446 passenger MESA 80 craft named the Seagull in 1979, while in the USA a SWATH workboat was built in 1972 called Kaimalino.

Trials with Kaimalino eventually led to building of four SWATH acoustic surveillance vessels for the US Navy, the Victorious class, at 71.5 m length and with a speed of 12 knots cruise speed (Fig. 6.39).

The typical SWATH configuration may be summarized as follows. Figure 6.33 shows the configuration of Duplus as an example;

- *Submerged body*: Torpedo like the submerged twin bodies with a cylindrical profile. Since the cylindrical profile simple and easy to configure, most of SWATH are of this type. Main propulsion engines (diesel or electric engines), and its transmissions to propellers, fuel tanks, ballast tanks, and remote control mechanism for stabilized fins, etc. are installed in the submerged bodies. The submerged bodies are made in steel or aluminium, and provide around 70–90% displacement.
- *Superstructure*: This is a box like structure for accommodating the working and passenger cabins, and generally made of aluminium alloy to minimize weight.
- *Struts*: The struts have a thin streamlined profile to reduce resistance and wave interference while piercing the surface to connect the submerged body to the super structure. In some craft, they are also used for power transmissions and crew access below.
- *Stabilizer fins*: SWATH generally have such fins mounted at the both bow and stern struts for improving transverse and longitudinal stability, as well as seakeeping, and controlling the dynamic running attitude.

The advantages of SWATH can be outlined as follows:

Minimized Motion of Ship in Waves

Since the wave disturbance to the craft is weak due to only the fine strut through the water surface, SWATH motions are much smaller compared with conventional craft, more like a semi-submersible drilling platform. The US Coast Guard carried out a comparative seakeeping test of the 220 t displacement "Kaimalino" (Fig. 6.34) with a monohull patrol ship weighing 3,100 t in sea state 3, and found that the motions of both craft were almost same, and even a little lower for the SWATH. The Japanese SWATH passenger ferry "Seagull 2" was operated in rough waves in the Japan Strait, and compared with conventional ships with similar displacement that the SWATH had an average rolling angle 1.5° at 24 knots compared to 9° for a monohull ship and vertical acceleration of 0.1 g compared to 0.6 g for monohull ship. The SWATH speed loss was only 2%, and with much less passenger seasickness.

Low Wave-Making Resistance

The wave-making resistance of such craft is very small; however, the body friction resistance is great due to the large wetted surface area (60% larger than conventional monohull) because of the large under water hull volume and deep draught.

Fig. 6.34 Kaimalino

The total resistance is larger than monohull craft at both low and high speed. Nevertheless, the total drag will be lower in the case of medium speed craft due to wave cancellation effects so there is a clear optimal operating point for SWATH. The SWATH vessels in Table 6.9 have Froude number clustered around 0.6–0.8. This has been found optimum for vessels with this type of configuration.

Large Deck Area and Superstructure Cabin Space

Similar to catamarans, the deck area and personnel accommodation or passenger cabins in the superstructure are able to be larger compared with conventional monohull ship, due to the large twin hull separation.

Manoeuvrability and Course Keeping

The SWATH profile with long twin under water hulls results in a craft with fine course keeping, and wide space between both propellers provides large turning moment even at low speed.

Propulsion Performance

Since SWATH draught is high, and hull separation is wide, the propeller diameter can be larger than that on conventional ships, so the efficiency can be maximized and cavitation avoided by SWATH. In addition, the under water hulls of SWATH usually are of regular slender cylindrical shape, so the wake around the propeller disc is more uniform, so improving hull efficiency. The total propulsion efficiency of SWATH can be 10–40% higher than a conventional monohull craft at its optimal Froude number.

Limitations of SWATH Craft

- *Deep draught*: this limits its application to deep waterways, harbours, and piers. The SWATH is an offshore rather than coastal vessel, which is natural when one considers the concept primary purpose is reduced response to a seaway.
- *Weight distribution sensitivity*: Since the buoyancy increase is very small as draft is increased in heave pitch or roll due to the small water plane area its stability is extremely sensitive to weight distribution and total payload variations. This not only influences the design and construction, i.e. more attention has to be paid on weight control during design and construction, but also influences the distribution of weight during operation, i.e. the variable weight has to be controlled more strictly than other craft.
- *Flooding Resistance*: The flooding resistance is poor compared with conventional ships due to the weight distribution issues mentioned above, particularly, in case of asymmetric flooding. For this reason, active ballast tanks and transfer systems have to be arranged on the ships for safety, so increasing ship light weight.
- *Less usable space*: Since the struts and lower hulls are too narrow to be usable as cabins and holds, the usable space within the hulls is rather low compared with conventional ships. This concept is useful for payloads that can be installed on the above water deck. The Victorious class is a typical example where the sonar and underwater listening gear is fixed equipment payload, the main changeable payload being the fuel and crew supplies.
- *Lower transportation efficiency*: The ship lightweight proportion of total displacement of SWATH is larger than conventional ships by up to 40%. Due to the large wetted surface and more complicated cross structure, lower hull, and struts as well as trim control system, power transmission, ballast system, etc. the SWATH is also more expensive to build than a catamaran. For this reason the transportation efficiency will be lower than a high-speed catamaran.
- *Lower speed*: Due to large wetted area and high friction resistance, the optimal speed for SWATH is between 10 and 25 knots (Table 6.9). In an attempt to improve on this a British company FBM developed a novel SWATH (Fig. 6.35)

Table 6.9 Leading particulars of example SWATH craft

Ship name	Duplus	Kaimalino	Marine ACE	Seagull	Kotozaki	Ohtori	Suavolino	Halcyon
Country	Holland	USA	Japan	Japan	Japan	Japan	USA	USA
Year	1969	1973	1978	1979	1980	1981	1981	1985
Builder	Boele	Curtis Bay	Mitsui	Mitsui	Mitsui	Mitsui	Poole Yard	RMI
Application	Working ship	Working ship	Test craft	Passenger Ferry	Waterway survey	Waterway survey	Test fishing boat	Test and Demonstration
Passengers	–	–	–	446+7	–	–	–	–
Displacement, tones	1,450	193	18.37	338	236	239	40	57
Loa, m	47	26.8	12.35	35.9	27.1	27.1	21.3	18.3
Boa, m	15.9	14.3	6.5	17.1	12.5	12.5	9.1	7.93
Draft, m	5.5	4.66	1.55	3.15	3.2	3.4	1.46	2.19
Speed, knots	8	25	18	27.1	20.5	20.6	20	21
Main engine	2×diesel	2×gas Turbine	2×petrol engine	2×diesel	2×diesel	2×diesel	2×diesel	2×diesel
Power, kW	1,260	3,265	298	6,010	2,820	2,820	625	757
Machinery location	Cross structure	Cross structure	Cross structure	Cross structure	Cross structure	Cross structure	Cross structure	Cross structure
Transmission	Electric	Chain drive	Bevel gears	Bevel gears	Bevel gear	Bevel gears	Bevel gear	Belt drive
Propeller	CPP	4b CPP	3b FPP	3b FPP	CPP	CPP	FPP	CPP
Stability fins	n/a	Automatic	4×Auto	4×Auto	Manual	Manual	Automatic	Automatic
Strut type	Single	Twin strut	Single	Single	Single	Single	Single	Single
Deck	Steel	Al	Al	Al	Al	Steel	Al	Al
Strut	Steel	Steel	Al	Al	Steel	Steel	Al	Al
Lower hull	Steel	Steel	Al	Al	Steel	Steel	Al	Al
Frl	0.234	0.79	0.85	0.74	0.65	0.65	0.71	0.81

Name	Chubasco	Seagull-2	Patria	Navitek 1	USS Victorious	Radisson diamond	Customs 201
Country	USA	Japan	UK	USA	USA	USA	China
Year	1987	1989	1989	1989	1991	1992	2001
Builder	James Betts	Mitsui	FBM, FDC400	Thomson metal	McDermott Shipyards	Rauma Repola	n/a
Application	Fishing boat	Passenger Ferry	Passenger Ferry, Madiera	Passenger Ferry	Navy anti submarine vessel	Passenger Cruise Liner	Customs working craft
Passengers + crew	9 + 3	410 + 7	400 + 10	450	5 + 19	354	n/a
Displacement, t	76	350	180	365	3,400	12,000	228
Loa, m	22.0	39.3	36.5	39.9	71.3	130	35.0
Boa, m	9.45	15.5	13.0	16.16	28.6	32	13.3
Draft, m	3.0	3.26	2.7	3.7	7.56	7.92	2.8
Speed, knots	20.0	27.5	32	15	10.4	12.5	17.5
Main engine	2 × diesel	4 × diesel	2 × MTU, 16V396TB84	2 × diesel	4 × diesel	4 × diesel	2 × diesel
Power, kW	1,113	7,620	2 × 2,040	964	1,187	11,345	2,240
Machinery location	Lower hull	Cross structure	Cross structure	Lower hull	Cross structure	Lower hull	Cross structure
Transmission	Directly	Bevel gears	Inclined shaft	Directly	Electric	Directly	Inclined shaft
Propeller	FPP 4b	2 FPP	2 FPP 3b	CPP	FPP 5b	CPP	FPP
Stability fins	Automatic	4 × Auto	None	Automatic	Automatic	Automatic	Fixed
Strut type	Single	Single	Single	Twin strut	Continuous	Continuous	Twin
Deck	Al	Al	Al	Al	Steel	Steel	Steel
Strut	Al	Al	Al	Steel	Steel	Steel	Steel
Lower hull	Al	Al	Al	Steel	Steel	Steel	Steel
Frl	0.70	0.72	0.82	0.39	0.20	0.18	0.49

Fig. 6.35 FBM Semi SWATH

with reduced under water volume of the lower torpedo hulls and improved geometry surface piercing continuous hull struts, to achieve a service speed of 30 knots, and with good seakeeping performance. Actually, this is a compromise between the high-speed catamaran, wave piercer, and classic SWATH. Similar basic ideas drove development of the Stena HSS craft and the Mitsui SSTH Ferry, so one may conclude that adopting a single hull strut and optimizing the lower hull buoyancy fraction can allow optimisation and higher service speeds.

- *Complicated power transmission*: In case the main engines are located on upper cross structure, which is the most traditional arrangement on SWATH as shown in Table 6.9, and propellers located behind the cylindrical hulls, a complicated Z type drive with bevel gears or inclined shaft drive with universal joints, or belt drive, or electric drive have to be installed on the ships. All of such arrangements makes the design more technically complicated and risky, and add more weight and cost. If the main engines are located in the lower hulls with direct power transmission, some of these problems can be avoided; however, the arrangement for engine removal for repair and maintenance will be made more complicated. Additionally, unless the power system in the hulls is electric, engine air intakes and exhausts all have to be lead through the struts. Ballast systems are also needed to keep an optimized running attitude at various loadings, maintain longitudinal stability and seaworthiness.

Applications

Based on the attributes discussed above the SWATH should be applied carefully where seaworthiness requirements are paramount and payloads can be close to static for example:

- Passenger ferry operating in rough seas, such as "Seagull-2" that operates in Japan. Figure 1.16 shows the general arrangement of this ferry ship, with length 35.9 m, beam 17.1 m, displacement 335 t, draft 3.2 m, speed 23.6 knots, 446 passengers.
- Naval Stealth missions; Figs. 6.36 and 6.37 show the US Navy Experimental Stealth Ship "Sea Shadow".
- Ocean survey ship, engineering research ship.
- Naval multiple purpose missions, such as submarine detection and surveillance, Fig. 6.39.

The sea slice, Fig. 6.38 was developed at the same time at Kaimalino, and while having high performance in a seaway, was rather sensitive to payload variations due to the very slender water plane of the struts. Navy research since then has shown that the continuous fine strut was almost as effective and could allow a more robust vessel to be developed. The Sea Shadow was one direction for such designs aimed at high speed, while for offshore patrol and survey a vessel class as shown in Fig. 6.39 was developed—the USS Victorious.

Fig. 6.36 USS sea shadow

Fig. 6.37 USS sea shadow

Fig. 6.38 Lockheed martin sea slice

Fig. 6.39 USS victorious class

Comparisons

We have looked at the different multihull configurations separately, so let us compare their characteristics and what we may expect as the next development steps with these configurations before we move on to hybrid concepts in the next chapter, as it may give some insights as to why the hybrids themselves have an attraction.

Initially, we have introduced the catamaran in its displacement form, which by utilizing a high L/B is able to achieve higher speeds than a monohull with similar powering at the cost of a more complex hull structure but not the complexities of other HPMV. To reach the speeds to compete with hydrofoils or ACV it is necessary to employ the wave piercing hull form, or planing hulls, or a combination of these. All of these have been developed by different shipbuilders and are now available in dimensions up to 120 m length, a size suitable for ocean crossing rather than just coastal navigation.

The motions of the catamaran base layout, whether wave piercing or planing do present a challenge in quartering seas. While the hull structure can be designed to cope with this, and freight payloads may not be sensitive, passengers and crew are a sensitive. Unkindly motions may cause sickness, but before that, people lose interest in food and merchandise on board, and these services are often important for earnings. The trimaran evolved by Austal provides a solution to this challenge for passenger service on exposed routes. The payload efficiency is nevertheless highest for the straight catamaran. It can therefore be understood why on the one hand the US Navy has selected the trimaran for the LCS with weapons payload more sensitive to

Fig. 6.40 Catamarans in Bergen

motions at speed, and the catamaran for logistic delivery across oceans. We discuss these programmes further in Chap. 8.

The SWATH has a niche market in rough water passenger ferry routes that are also relatively short so that speed is not such a factor. The high performance here is operation in a high seaway without loss of speed, rather than high speed as such. Leading particulars of some example SWATH are listed in Table 6.9 above.

It has been mentioned that craft such as WPC utilize stabilizer surfaces, if we take this one step further and use these surfaces to support all or part of the craft weight we move into the region of hybrid concepts.

First a thought on where do catamarans go from here? The racing and record breaking craft have hit a performance boundary a few times in the last half century. The key at present seems to be increased stability at the edge of their performance. The assistance of computer technology will clearly move the boundary upwards bit by bit, though at the speeds they already operate, it is rather like the progress made by Formula 1 car racing teams, mostly only visible to the aficionados themselves rather than the public.

Commercial catamaran and perhaps also trimaran development is most likely to be focused on the power system, as today's high-speed diesels are developed further, to burn gas (from LNG or CNG in special tanks); to cleaner and more efficient fuel burn in the cylinders, and in the medium term future possibly to move to fuel cell and electric power at least for smaller passenger ferries once the total efficiency and power level of these systems is high enough and large enough.

In the meantime, the worldwide fleet continues to grow and be refreshed with replacement ferries of increasing comfort and efficiency. Figure 6.40 above shows two craft in Bergen moored overnight in March 2011.

Chapter 7
Novel and Hybrid High-Speed Craft

Hybrid HPMV

The last half century has seen a focus on development of HPMV having all or most of their support from one means—airfoils, hydrofoils, pumped or dynamically generated air cushion or planing forces from the water surface itself, as we have discussed in the previous chapters.

Hull geometry has also developed, mainly following variations of the multihull format for HPMV, in an attempt to improve seakeeping performance. Increasing physical size has helped this moving from waterline lengths of 20–35 m up to as much as 120 m for the largest catamarans and trimarans presently. The various hull forms have shown their limits, particularly in terms of dynamic stability in a seaway, and that has caused shipbuilders to add stabilizer fins and stern flaps or flow interrupters to monohulls, and both these together with forward-mounted bow stabilizer hydrofoils to catamarans and trimarans.

These devices have been quite successful to reduce roll and pitch response in a seaway. The obvious question is whether these devices could be used to support craft weight and so help reduce resistance as well. Second, what if we mix and match air cushions with hulls to remove water contact with part of the hull surface, or modify hull geometry to provide a much more stable dynamic air cushion support? Since the early 1990s, shipbuilders and operators have begun to experiment with a number of these hybrid concepts as a form of optimization, with increasing success. We'll introduce a few of these concepts here and talk a little about how we see the future prospects for them.

High-speed craft are supported by varying proportions of four kinds of lift force, i.e. Static Buoyancy, Hydrodynamic Lift, Aerodynamic Lift and Air Cushion Lift. One of the best ways to improve overall performance is by using combinations of the various supporting forces, and according to particular craft design requirements, to use these combinations to optimize the balance of lift and drag forces so as to smooth the takeoff drag hump, and then give high seakeeping performance.

L. Yun and A. Bliault, *High Performance Marine Vessels*,
DOI 10.1007/978-1-4614-0869-7_7, © Springer Science+Business Media, LLC 2012

In previous chapters, we have already introduced two hybrid HPMV:

- Catamaran + Planing Hull = Planing Catamaran, (PCAT)
- Wing in Ground Effect + Air Cushion = Dynamic Air Cushion WIG Craft (DACWIG)

In this chapter, we will introduce some more hybrid craft:

- Planing monohull craft + Air Cushion Cavity = Air Cavity Craft (ACC)
- Catamaran + Hydrofoil = Foil-Assisted Catamaran (FACAT)
- SWATH craft + Hydrofoil = Hydrofoil SWATH Craft (HYSWAC)
- Semi-SWATH CAT with Bulbous Bow
- Air Cushion + Catamaran = Partial Air Cushion-Supported Catamaran (PACSCAT)
- M Craft

Air Cavity Craft

ACC are quite different from the ACV as the design intent is not to create a continuous air gap under the hull. The ACC concept originates from the high-speed planing craft with transverse steps, and a thin air layer attaching to the bottom of the craft that isolates part of the immersed hull surfaces from the water so as to reduce skin friction forces [7-1,7-2,7-3]. The ACC takes an approach similar to the SES, while having a shallower "cavity" underneath the hull and more substantial "side-hulls" or simply a cavity "cut out" of the underside of the hull. The hull structure is therefore less slender and can be built at larger sizes more efficiently. The downside is that such craft require higher power levels for similar speed at a given size compared to SES. The upside is that they could be larger! The key to success is the design of the cavity, remembering that at sub-planing speeds the water surface in the cavity will be an irregular surface due to two factors—the pressure variations as surface waves pass the ship hull outside, and second, the generation of waves from the front of the cavity. These waves will pass though the cavity and will vary with speed, eventually smoothing out once planing occurs if the craft can reach this speed. Once planing, the air fed into the cavity has to have sufficient volume and static pressure to maintain cavity depth all the way to the stern, as otherwise the cavity will operate like the step in a stepped planing hull and fade away over a short length.

Figure 4.9 shows a profile of a high-speed planing craft with transverse steps aft of amidships. When running at high speed, the wetted surfaces of the craft are just forward of amidships and close to the stern on the bottom, as described in Chap. 4. At this speed, water friction will be a large percentage (around 70–80%) of the total drag. Designers sometimes try to induce some airflow just behind the step. Low air pressure or even suction due to water vapour pressure is available here, and this can reduce the total pressure needed to be supplied by fans, if mechanical blowing from the step face were used rather than depend on cavitation vapour pressure and air sucked in from the side of the step. Having a wedge form to the centre structure

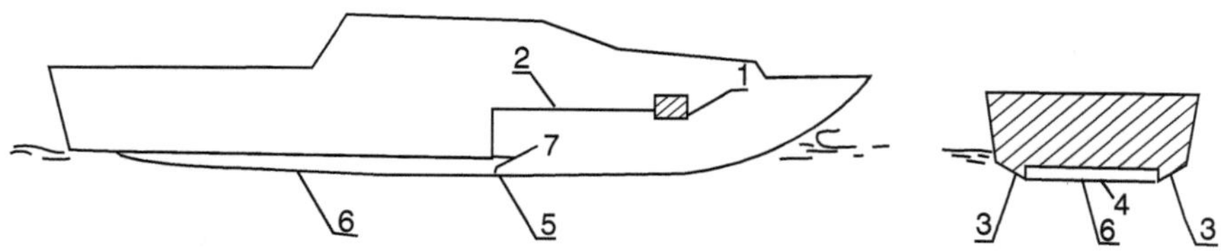

Fig. 7.1 Air cavity craft (ACC) system diagram

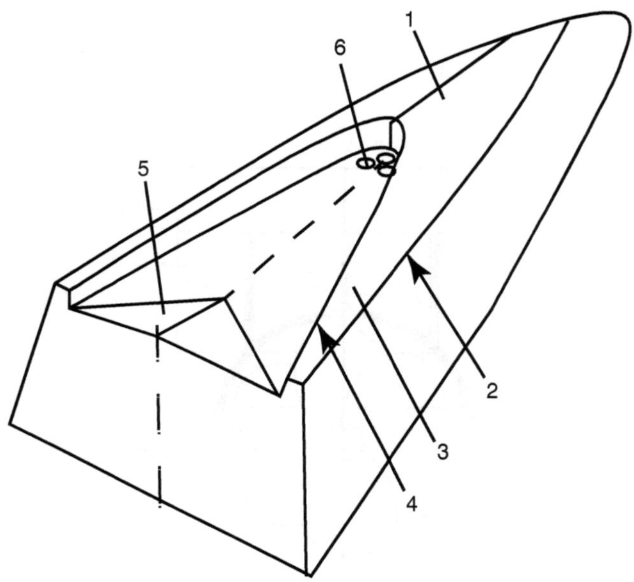

Fig. 7.2 3D Model of an air cavity hull seen from underside

underside of a catamaran can also assist generating dynamic pressure. The aim is to generate an air/vapour layer covering the "wetted" surface of the planing surface behind the step and reduce the resistance of this part, since the density of air is only one 800th of the water. The ACC extends this approach as implemented on stepped monohulls and planing catamarans by introducing the central pressurized air cavity.

Figure 7.1 shows a diagram of an ACC, where 1-air compressor; 2-air duct; 3-side keel or skeg; 4-central longitudinal keel (for improving transverse stability); 5-transverse step; 6-air layer.

Figure 7.2 shows a typical 3D model of an ACC hull, inverted, where 1-forward hull bottom; 2-hard chine; 3-skeg or side keel; 4-edge of air cavity; 5-planing stern part of bottom; 6-holes for feeding compressed air to the cavity.

Figure 7.2 has the typical configuration of a planing ACC, which includes significant deadrise at the bow, one for improving seakeeping quality, a flattening

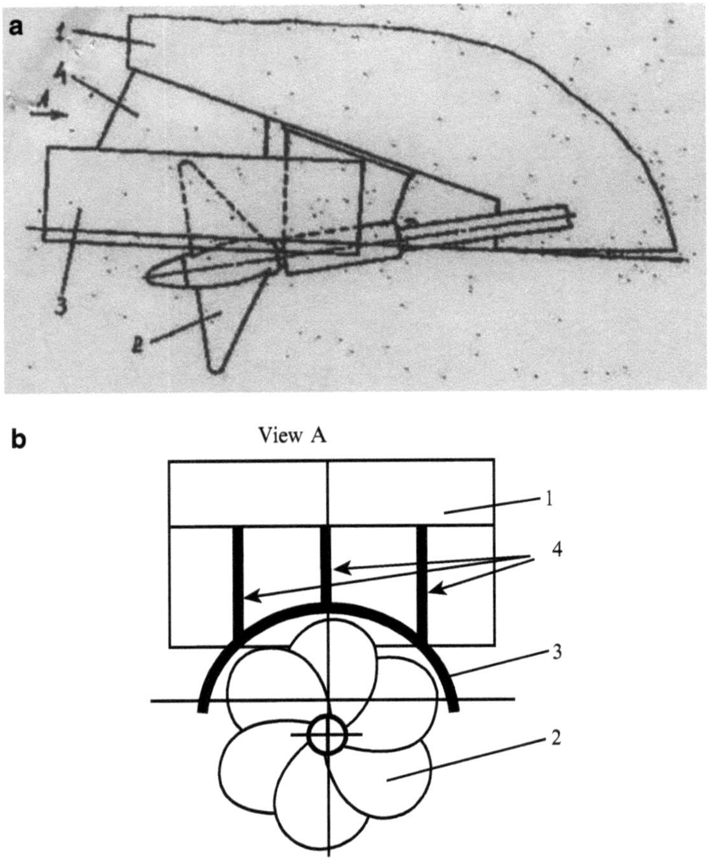

Fig. 7.3 (a, b) End and side views of surface propeller systems for ACC HPMV

deadrise of the hull bottom outside the air cavity for planing at high speed (2, 3, 4) and a reverse-angled wedge at the stern which helps retain the air/vapour cushion.

A most important issue in ACC research has been how to resolve the problem concerned with thrust and efficiency decrease due to the suction of air into a marine propeller disc. Since the air layer is attached to the planing surface at the stern, a water propeller arranged at the stern under the water surface can be affected by air ventilation as well as cavitation, and consequently experience reduced absolute thrust and propeller efficiency.

One approach is to isolate the air from the propeller by using a fence or tunnel form, as shown in Fig. 7.3, in which (a) the profile of stern part of the craft with air protection equipment; (b) shows the transverse section of the stern part of the craft. The items in both figures are 1-craft hull; 2-propeller; 3-semi-tunnel; 4-semi-tunnel support brackets.

Key Features of ACC may be summarized as follows:

Low total drag: The friction drag can be decreased as much as 70–80% when planing, and total drag decreased about 20–25% compared with conventional planing craft, due to positive air lubrication of much more of the hull lower surface than is normal for a stepped planing hull.

Low lift power: Since the air layer in the hull cavity is very thin, and the exhaust is only at the stern, airflow can be low, so fans absorbing only 2–3% of total power output of the craft are required. This can be compared with skirted air cushion craft where the cushion typically takes more than 25% for ACV, and 15% for SES.

Low impact loads: The impact (wave slamming) load on the hull may be reduced by up to 50%, due to the air layer on the bottom, compared with the conventional planing craft. This is a significant factor influencing hull design and gives improvements to passenger and crew comfort and equipment vibration, as well as reduced speed loss at high speed in a seaway. It must be noted though that the cavity must be carefully designed so that waves inside the cavity are suppressed by the fan pressure and flow sufficiently to avoid cavity blockage and consequent increased hull resistance, as below.

ACC key design challenges are as follows:

Loss of air layer: Since the air layer is thin, it can also be lost when running in rough seas with heavy motion. The ACC is therefore better suited to coastal and estuarine conditions rather than deep sea routes. This was taken advantage of in Russia in the 1970s and 1980s where a number of craft were built for passenger ferry application—a simpler alternative to the shallow submerged hydrofoil craft.

Performance in a seaway: The ACC is limited in the wave height that it can operate due to the dynamics of the air cavity. This presents clear limits to the environment an ACC craft can be designed for as it is not practical to deepen the cavity unless the configuration is changed radically—effectively to a catamaran with a stiff bow seal and an air lubricated stern seal between the side hulls. This configuration has in fact been developed in the UK and is under trials in the form a craft called PACSCAT which we discuss further below.

We show an example of a Russian ACC Ferry in Fig. 7.4, while Fig. 7.5 shows a Russian ACC landing craft for river and inland sea operation. Figure 7.4 shows the passenger ACC the "Linda", completed in the 1992, where (a) shows the craft in operation; and (b) general arrangement of the craft. Ten craft have been built to the same design for service on the Russian river system. Figure 7.5 shows a landing craft ACC Serna also constructed in Russia, where (a) shows the craft and (b) shows the general arrangement. Five of these craft were built in series in the 1990s, and several other craft since then, see [7-4].

Table 7.1 on page 263 below gives key data on some example ACC craft produced in Russia.

This type of craft has had a resurgence of interest in the last few years particularly in Norway where a new prototype has been built. Effect Ships International

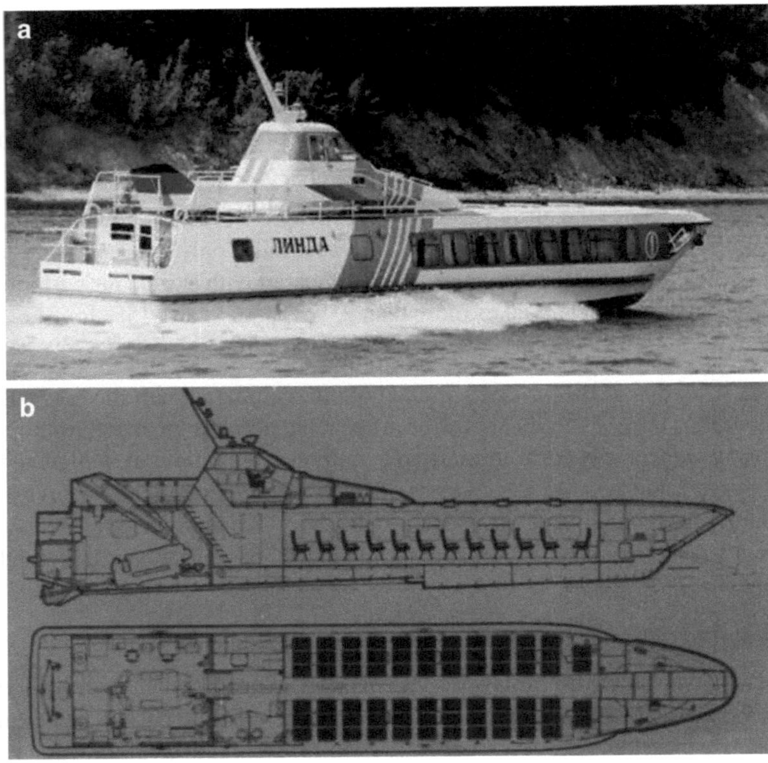

Fig. 7.4 (**a**) ACC Linda photo. (**b**) Linda general arrangement

has built a 10-m catamaran test craft; Fig. 7.6a to verify its patented cavity geometry. Model Test of a 40-m catamaran geometry seen from under water is shown in Fig. 7.6b, at 70 knots. One can see the propulsion gondola centre stern in each hull, and the fine dart-shaped cavities. Moving to full-scale ESI has first built a 20-m test monohull that can exceed 30 knots powered by a modest 640 kW [7-5] supported by funds from a European Union technology programme. The craft is shown in Fig. 7.7a. This has a rather fuller mid-ships profile than the catamaran in Fig. 7.6b, suited to lower service speeds.

The ESI cavity form is deeper than normal air lubricated vessels, with the surrounding structure much more like the SES form than the Russian ACC craft above. Air cushion support is aimed at up to 85% of total weight, which is as high as most wide side hull SES and allows the planing surfaces to be diminished. The challenge is the propulsion and its separation from the air cushion. ESI are developing water jets installed into internal gondola's fine hull shaped forms at the stern quarter of the craft with emersion for the intakes set low enough that ventilation is not a

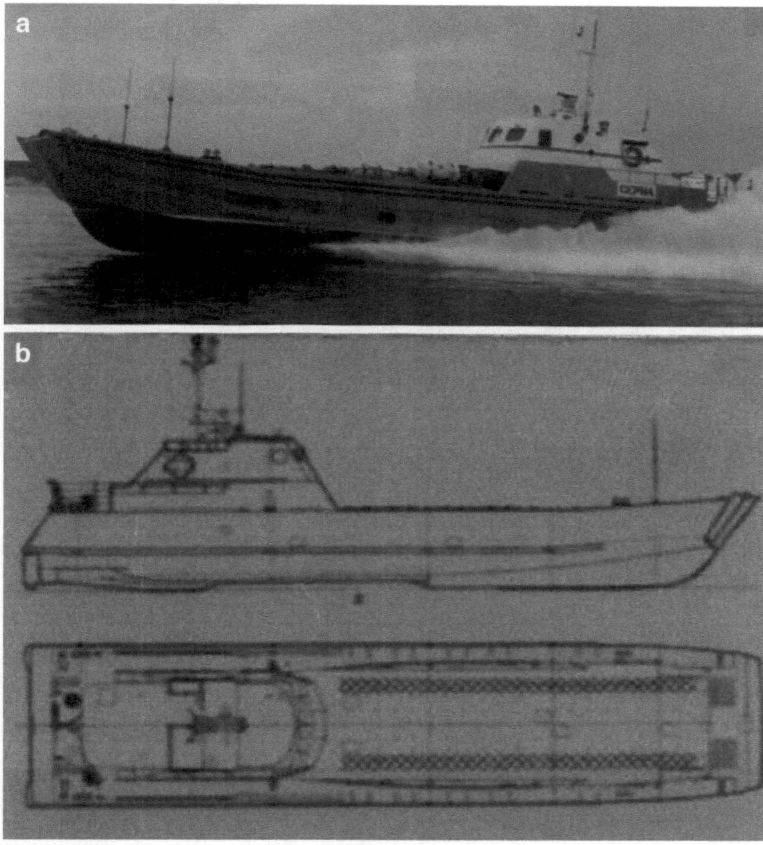

Fig. 7.5 (**a**) ACC landing craft. (**b**) General arrangement

Table 7.1 Leading particulars for some Russian-operated air cavity craft

Name	Serna	Linda	Mercury	Tornado
Delivery year	1992	1992	1995	2000
No of ships	5	11	4	2
Displacement (t)	105	24.6	99.0	30.8
Loa (m)	25.65	24.1	35.4	19.6
Boa (m)	5.85	4.6	8.3	3.9
Max draft (m)	1.52	0.95	2.0	0.9
Power (kW)	2×2,430	1×660	2×3,670	2×1,220
Max speed (kt)	32	38	52	50
Range (nm)	100	220	n/a	n/a
Passengers	75, 45 t	70		
Mission	Landing craft	Fast ferry	Customs patrol boat	Fast patrol boat
Propulsion	Ventilated water jet	Semi-submerged propeller	Propeller	Semi-submerged propeller
Frl	1.04	1.27	1.44	1.86

Fig. 7.6 (**a**) ESI ASV Cat test prototype. (**b**) ESI 40-m Cat underside view

problem in rougher conditions. Results from the ESI monohull prototype, Fig. 7.7b, show a significant reduction in fuel consumption at equivalent speed to planing monohull craft.

The ESI hull geometry is complex. Fabrication in fibre-reinforced resin in a mould is probably a cost-effective method available for geometry such as this. Fortunately, the technology for such moulded hull construction has become both advanced and cost-effective for passenger ferry construction over the last decade, so that a number of shipyards are now building catamaran ferries in carbon fibre-reinforced material. Extension to utilizing ESI's air-supported hull concept may enable reduction in power installed for a given service speed for fast coastal patrol craft.

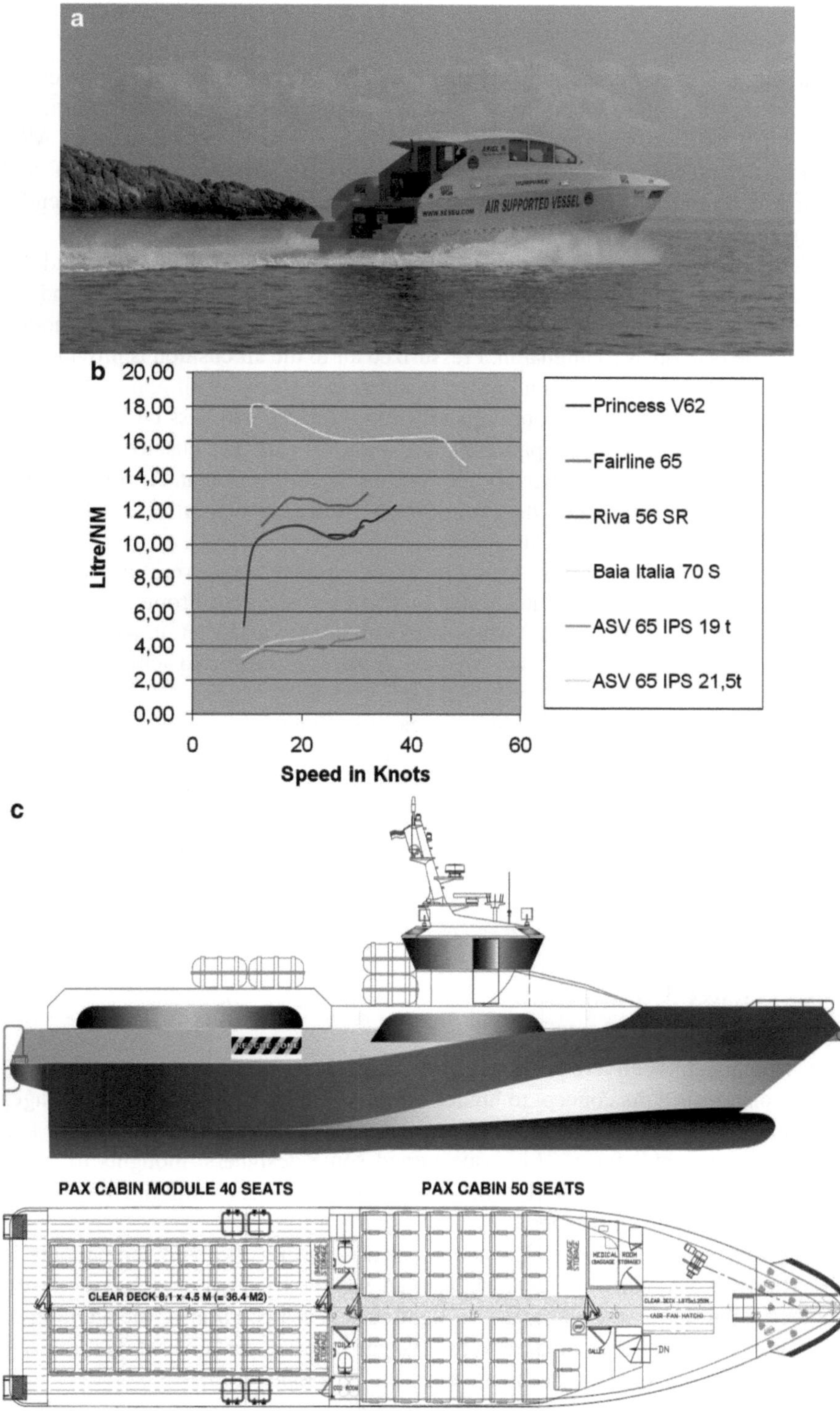

Fig. 7.7 (**a**) ESI ASV 20-m monohull. (**b**) ESI test results. (**c**) ESI ASV crewboat general arrangement

In fact, it is the offshore industry that has taken the concept on for commercial application—a fast crew boat to be built at Indonesian Orela shipyard, Surabaya for launch early 2012. The 22-m vessel is being built in aluminium, an ASV Monohull form with a beam of approximately 5.15 m. To reduce pitch motions and minimize slamming, the ASV features a sharp and flared bow with a slender extended bulb, Fig. 7.7c.

At the stern on either side of the air cushion cavity, there are propulsion gondolas housing water jet intakes. Dynamic lift is secured from planing surfaces around the vessel's LCG as well as providing buoyancy. This configuration reduces roll motion in the same fashion as a catamaran. Pressurized air to the air cushion is provided by a hydraulically operated lift fan system located in the bow, well above the water line. The air cushion covers approximately 46 m^2 of area with less cushion surface per hull length in the forward part and more in the aft part. The cushion is enclosed in the aft section with a proprietary air cushion flap arrangement. At full load, the cushion pressure is 6–7 kPa depending on the selected support rate, which can exceed 80% of the vessel displacement.

The vessel trim is very close to neutral at design speed; however, for fine-tuning the trim angle, a Humphree Interceptor system may be fitted. Propulsion consists of twin Rolls Royce 45 Series 3 water jets, with unique water jet intake geometry. With a propulsion power of 2×800 kW, the vessel is estimated to achieve 40 knots at full load.

The first vessel will have seating capacity for 50 passengers on the main deck. Behind the cabin area, there will be a large deck area of almost 40 m^2 to accomodate light cargo, container modules of different kinds or as showing on the general arrangement, a second passenger cabin module with additional 40 seats. An additional space is allocated for crew accommodation, toilets, medical room, galley, baggage storage, etc.

Foil-Assisted Catamaran

The catamaran has delivered high performance in the medium speed range of operation. If one pushes the concept to higher speed, with planing hulls, then challenges arrive, unless you are willing to accept the higher accelerations of the planing and stepped planing hull form. This issue turned some designers' thoughts to see if hydrofoils could assist to stabilize motions at high speed and combine the positive attributes of the two craft types.

In recent years, a number of research and design institutes and shipbuilders in UK, Russia, USA, Norway, Japan, China and South Africa have worked on development of foil-assisted catamarans. In light coastal sea conditions, the addition of foil support for smaller catamaran ferries has indeed been proved useful to improve efficiency and ride quality, and so a market has developed for these.

As a reminder of the characteristics of the separate types, we summarize advantages and disadvantages of both CAT and HYC in the table.

CAT and HYC Compared

	CAT	HYC
Advantages	• Wide cabin area • High transverse stability • Simple structures	• High lift/drag ratio at high speed • Fine seaworthiness with automated control system
Disadvantages	• Large drag at high speed • Lower seaworthiness, with higher seasickness rate	• Small deck area • Controllability issues in case of automated control system failure • High cost with complicated technology in case of fully automated foil control

Figure 7.8 shows a FACAT type "Superfoil-40", designed and constructed in Russia [7-6] and operated in the Baltic Sea between Helsinki in Finland and Tallinn in Estonia from 2000 to 2004. Figure 7.9 shows the general arrangement of the aluminium hulled craft. In this case, foils are used to raise the bow of the catamaran out of the water allowing it to plane on the stern part of the hulls only.

Figure 7.10 shows a frontal view of the craft static with twin bow foils retracted. The initial concept as for propulsion by open propellers as can be seen in the General arrangement. To improve performance, the craft was to be fitted with stern interceptors. Figure 7.11 shows a 2D schematic of the interceptor hydrodynamics [7-7], and one can see the lift increase due to the interceptor. The leading particulars of "Superfoil-40" are Loa 41 m, Boa 12.4 m, Max draft foils down 3.4 m, foils retracted 1.7 m, craft running on foils 1.2 m. The craft as built has a maximum speed of 55 knots driven by 4×MTU 12V4000M70 high-speed diesels of 1,740 kW each, driving 4 MJP water jets. It can accommodate 286 passengers.

Fig. 7.8 Foil-assisted catamaran (FACAT) Superfoil 40

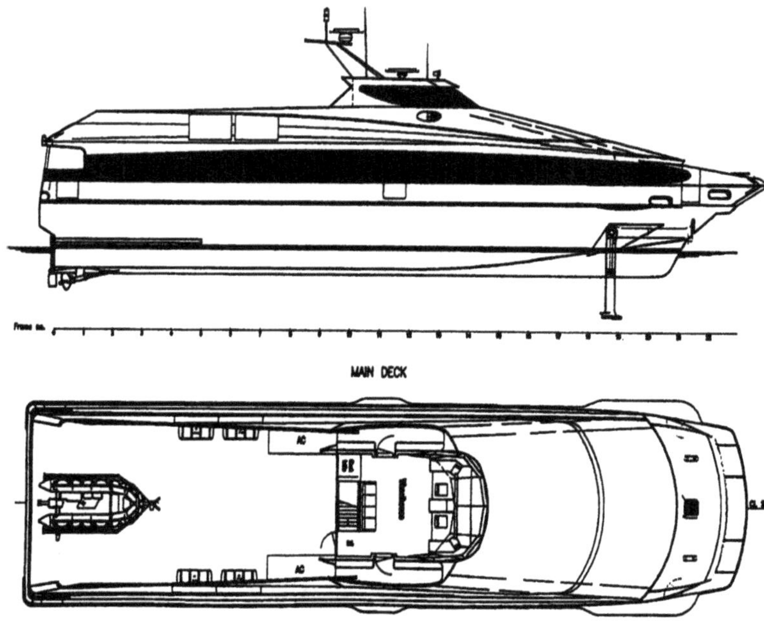

Fig. 7.9 FACAT superfoil general arrangement—initial design

Fig. 7.10 Foils retracted front view

The two retractable foils arranged at the bow under the twin hulls support about half of craft weight, and the twin hulls bottom at the stern is designed as flat and wide so as to support planing at high speed with minimized wetted length, supporting about half of craft weight. Most of the twin hull length is lifted out of the water surface at high speed. The interceptor is a vertical plate moving automatically in

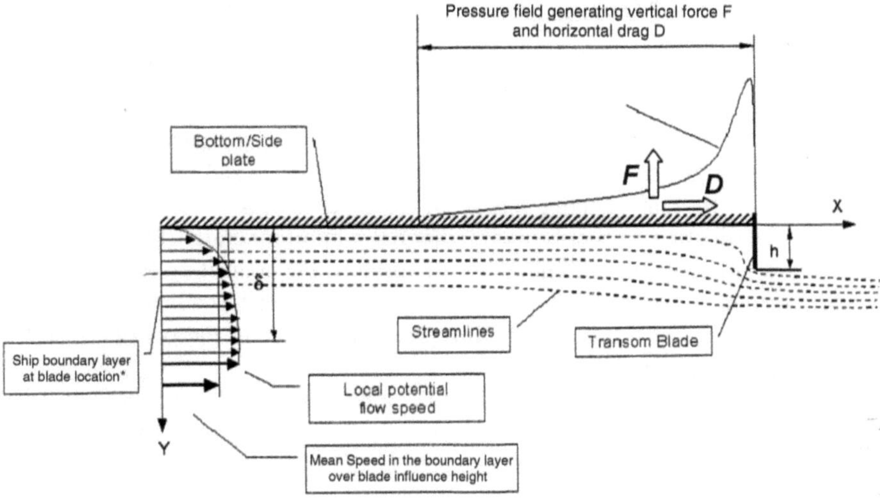

Fig. 7.11 Interceptor schematic

vertical direction, a small distance protruding below the stern transom to adjust the hydrodynamic pressure and lift acting on the stern bottom of the twin hulls for improving the seakeeping quality and takeoff performance.

The stern interceptors protrude only slightly under the transom. The height of interceptor below the transom is about 0.1–0.2% of craft length; however, since they change the pressure field forward of the transom, they increase the lift under the bottom in the stern region, thus decreasing the wetted length and hull friction drag. In addition, as the interceptor height is so small, less than the thickness of the turbulent boundary layer under the stern, the drag due to the "roughness" is small.

Use of an interceptor therefore gives increased lift to the twin hulls and decreased drag, due to decrease of hull wetted length, thus increased hydrodynamic efficiency. These devices have also been used on fast monohulls as well as on catamarans. Raising and lowering is simply by hydraulic jacks so the concept can be applied simply in an automated control system for improving seaworthiness as well as the power performance. Their effectiveness is similar to a moveable stern flap. The combination of foils and stern interceptor is very effective at high speed giving a higher performance than HSCAT, and with shallower draft. On the FACAT, a simple automated control system is installed to adjust the bow foils flap angle and interceptors at stern transom so as to maintain steady trim.

When running in shallow water at low speed (before takeoff), the bow foils can be retracted to reduce the draft of the craft from 3.4 m down to 1.7 m, so as to enhance craft operability.

The advantages of the FACAT can be summarized as:

- *High speed*: up to 50 knots, compared to usually 35 knots for CAT, with high hydrodynamic efficiency.

- *Spacious payload deck area* and cabin on the upper deck similar to a catamaran:
- *Seaworthiness*: Seaworthiness improves significantly compared with the basic CAT of the same configuration when an automated control system is installed to adjust the flap angle on the foil and the movement of interceptors, minimizing speed loss to just 10–15%, vertical acceleration of 0.1 g and passenger seasickness of zero for craft running in head waves in sea state 4.
- *Stability*: With bow foils only, used for partial support for craft lift from the water rather than maintaining flight as for a hydrofoil, the control system can be far simpler than that on a fully submerged hydrofoil, giving cost advantage. In addition, high inherent transverse stability is available due to the twin hull, so the craft has low roll motion
- *Easy mooring alongside jetties*: Due to the foils being installed between the twin hulls, without protruding outboard each side, the craft is more convenient to be moored than a conventional HYC with surface-piercing foil systems

The limitations of the FACAT with bow only foil configuration are that the sea state suitable for its operation is set by the submergence of the bow foils. In rivers, lakes and sheltered coastal conditions, it can perform well, while in open seas the bow foils will surface in wave troughs and performance will degrade towards the basic catamaran.

If the foil-assisted catamaran is to have improved performance, then a fully submerged foil system at both bow and stern has to be designed—this is the challenge addressed by Fjellstrand in the early 1990s with the Foilcat, and also Westamarin with their version. We have introduced the Foilcat at the end of Chap. 5 as it is really an extension of the hydrofoil story. Fjellstrand were successful with their development and eventually built a number of such craft that are now in service for Turbojet on the route between Hong Kong and Macau. We show one of these craft in Fig. 5.36.

Foil-Assisted SWATH (HYSWAC)

The advantage of a SWATH is its very high seaworthiness; however, due to the large wetted area of the twin hulls, the friction drag is high, and consequently, it is only useful for medium speed craft, say around 18–25 knots.

A novel hybrid of the SWATH and hydrofoil, the HYSWAC, was developed by Navatek for the Office of Naval Research in the USA in 2002. The US Navy SES 200 that was developed from the BH110 design of Bell Halter for trials of a diesel-powered medium speed craft for coastal patrol and logistics was converted to this configuration in 2002–2003. After trials with Navatek in 2003, it was tested by the US Navy in 2004–2005 in Hawaiian waters. The foil configuration is shown in Fig. 7.12. At the craft amidships is the large submerged lifting body that encloses right angle gearboxes and horizontal shafts out to the twin propellers. At the stern is a horizontal lifting foil with control flaps on each side, and single central rudder.

Fig. 7.12 Sea Flyer foils

Fig. 7.13 Sea Flyer buoyant body

The craft has a displacement of between 274 and 325 t depending on ballast condition, length 48.78 m, beam 13.1 m draft 5.6 m. It is powered by two 4,000 kW diesel engines, each mounted inside the main hulls driving propellers via two right angle gearboxes at deck and in the lifting body, giving a service speed of 30 knots. Figure 7.13 shows a view of the main lift body located near to amidships. The lifting body has a length across the vessel of 10 m giving a displacement of 163 t, equal to about half of the craft displacement, and is used in the same way as the two cylindrical buoyant bodies of a normal catamaran SWATH, providing large added mass and

Fig. 7.14 Sea Flyer at speed

damping force restraining the craft motion in waves. In this case, the buoyant body is shaped as a hydrofoil and generates lift force at high speed in addition. With the aid of an automated control system, flaps at the trailing edge of the submerged foil can be controlled to further improve the motions in waves. The craft, renamed Sea Flyer at speed, is shown in Fig. 7.14. Sea trials in seastates up to 3.5 m seas showed the vessel to be very stable [7-8].

Navatek have continued their development of submerged body-enhanced vessels with the HDV-100 "blended wing and lifting body" modified monohull craft, shown in Fig. 7.15a, b. The 30-m craft is used as a technology test bed for Navatek. The company has also worked on SWATH craft, as well as a number of performance-enhancing variations to the deep vee planing monohull, see their web site in resources at the end of this book.

The resulting key attributes of the HYSWAC craft can be outlined as:

- *Seaworthiness*: Thanks to the main submerged buoyant body, the craft has fine seaworthiness at low speed, like a SWATH (with low wave interference, high motion damping coefficient and large added mass), and also fine seaworthiness at high speed, similar to an fully submerged hydrofoil catamaran, where the hulls emerge above the water surface to reduce both wave-making and friction drag, and also wave interference.
- *Long range*: Thanks to the large volume lift body available to accommodate a large amount of fuel, the range is high compared to many other fast craft
- *Medium speed*: The craft can be operated as high as 30 knots, much higher than a conventional SWATH, and it will be higher still if craft size is further increased.

Fig. 7.15 (a) HDV-100 foil arrangement. (b) HDV-100 at speed

Semi-SWATH CAT with Bulbous Bow

The advantages of the catamaran are its large deck and cabin space for accommodating payload and passengers, high stability and a relatively simple hull structure compared with other concepts. Its disadvantage, however, is high drag and so high power demand at service speed in waves, combined with significant motion response in cross seas.

To improve on the basic configuration, a CAT with finer water plane lines and bulbous bow was developed by Stena Shipping together with Rauma Repola in Finland at the beginning of the 1990s. The craft was called "HSS" for high-speed ship and was aimed at car/truck/passenger traffic between Holland, the UK and Ireland. The HSS has been discussed in Chap. 6. More recently, a fast catamaran "semi SWATH" with bulbous bows for navy use was developed by the BMT Company in UK in 2005, called the "X" craft [7-9]. This was a design for the US

Fig. 7.16 BMT X craft for US Navy

Navy Littoral Combat Ship (LCS) competition. The X craft mixes the SWATH and CAT hull forms.

The semi-SWATH is a ship form in between the conventional CAT and the SWATH. The design water line is constricted and the centre of displaced volume is moved further below the design water line so as to give finer lines at the surface but without increasing the draft. There is a sharp bulbous bow to reduce wave generation. The bulbous bow serves the normal purpose of wave cancellation to minimize wave making drag. The combination of finer waterline and bulbous bow optimized to the bow shape also improves the seakeeping. The semi-swath cross-section looks a little like a typical ancient Grecian Amphora Vase.

In general, there are two sub-types of Semi-SWATH as follows:

- Water plane wholly constricted—the Stena HSS760 and 1,500. On these craft, the water plane is constricted for the full length, with a small and sharp bulbous bow, as shown in Figs. 6.7 and 6.8.
- Water plane partly constricted—the "X" craft shown in Fig. 7.16 that has been developed by BMT in the UK. The water plane is constricted mainly forward of amidships and at the bow, with a small bulbous bow and an almost conventional catamaran shape aft of amidships. Figure 7.17 shows the craft at speed.

Features of the Semi-SWATH

The effect of a bulbous bow to improve both transverse and divergent wave interference so as to reduce craft wave making resistance is very weak in the case of a normal high-speed catamaran at high Froude number. Its effect is reduced by the high hull slenderness, as well as by large hull separation. In addition, a small reduction of wave making cannot compensate for the increase of wetted area of the bulbous bow and associated higher friction resistance.

Fig. 7.17 Sea Fighter at speed

By moving most of the displaced volume below the design water plane forward of amidships and using a finer form at the water line towards the bow, a sharp bow and stem is formed at the design water line, thus a small and sharp bulbous bow may be added which will have a similar effect to the wave piercing form described in Chap. 6. The principal objective with the semi-SWATH is also similar to that for the wavepiercer—improved seaway performance.

Improvements in seakeeping for Semi-SWATH are as follows:

- Significant decrease of heaving amplitude, due to decreased wave perturbation in a seaway and increased natural heave period, as well as increased heave damping force from the submerged hull shape
- Decreased vertical acceleration and seasickness of crews and passengers, due to the increased pitching period, and damping moment, similar to the heave response improvements
- Decreased speed loss in waves, due to the decreased motion and sharp bulbous bow

The differences between the whole length Semi-SWATH (HSS1,500/2,100) and part Semi-SWATH (X craft) are the S type body extended to the whole length of the craft to improve the seaworthiness for the HSS, while the X Craft has finer lines for the forward part of the craft. The X craft employs additional hydrodynamic damping in the form of hydrofoils beneath the bow and an interceptor system at the stern to improve the seakeeping as follows:

- *Submerged Bow hydrofoil*: A submerged bow foil located at a quarter of the length from the bow, with controllable flap for improving seakeeping;
- *Stern interceptor*: The stern interceptors are located at the transoms of the twin hulls, similar to the FACAT described above, to improve seakeeping by acting together with the submerged bow foils and also to improve the running attitude and optimize the drag.

The leading particulars of the "Sea Fighter" are Loa 79.9 m, Boa 21 m, Draft 3.6 m, Max Speed with combat load 50+ knots, in SS4 40+ knots, Cruise on diesels 20+ knots, Range at cruise 4,000 nautical miles, accommodation 50, displacement 1,400 t, payload 500 t. Main machinery consists of 2 × GE LM2500 gas turbines of 23,380 kw each and 2 × MTU16V595TE90 diesels 4,350 kW. The engines are coupled to 4 KaMeWa 125 SII water jets. The vessel has deck landing pads for 2 × SH-60 or equivalent helicopters.

The Sea Fighter was designed to accept changeable containerized mission support packages so as to be tested for a range of possible missions. The design and build contract was let in February 2003. It was designed by BMT Nigel Gee, built at Nichols Brothers, Washington, and launched 2 years later in February 2005 followed by delivery to the US Navy July 1st that year after builders trials. Inside the craft is installed an X-Y gantry crane system that can pick up and place up to 12 ISO 20 foot mission modules. Each of the locations was designed to hook up to power, phone, LAN and chill water. At the stern is installed a combined RoRo loading ramp and RIB boat launch ramp that is deployed hydraulically.

The vessel met its speed specification very well, reaching up to 54.6 knots top speed in calm conditions. The ride control system and steering system with water jet nozzles and active skegs at the stern gave tight manoeuvrability for a craft of its size. Teething problems were experienced, including system integration issues and need to upgrade the telecoms mast that was implemented within 8 months. Such issues are to be expected in a prototype vessel, and the US Navy programme accounted for resolving them. A full scientific data-gathering package was installed to monitor craft performance during the performance trials carried out from October 2005 through mid-2006; see [7-10].

We discuss the US Navy LCS and JHSV programmes in the next chapter. What is clear from the experience of the trials with SeaFighter is that it is a very capable craft and was able to be used to test important parts of both the missions. Once there is a reliable helicopter operational base, a hangar and service facility is clearly valuable, which has been adopted on the LCS, the containerized operational modules can be very useful, while the hydraulic loading ramp has shown itself as a valuable attribute for JHSV operations to many austere or degraded quayside locations. The combination of semi-swath hull form with high power water jets has produced a very fast ship. The subsequent specification for the operational programmes has been selected lower than this, staying well within documented performance data—a conservative approach.

Partial Air Cushion-Supported Catamaran

The PACSCAT was previously called the Air Cushion CAT. The concept is a craft with high length/beam ratio based on a catamaran configuration and medium cushion pressure to reduce the resistance at hump speed. Resistance is also improved at cruising service speed, as the air cushion support reduces effective draught by reducing the hull wetted surface on the inside and so improves hydrodynamic efficiency compared to a catamaran, at the expense of the power needed for the lift system.

The PACSCAT is a hybrid of the ACV and CAT that uses a lower cushion pressure than needed to support the total weight of the craft. The internal cushion itself meanwhile is deep, so responding like the cushion of an SES rather than the shallow cushion of an ACC. A cross-section of PACSCAT is shown in Fig. 7.18. The side hulls are wider and more buoyant than those typical of SES designs. Diesel engines are used to drive the lift fans and the propulsion water jets in each side hull. The side hulls widen significantly at the stern, providing space for this machinery and reducing the escape area for the cushion air at the stern. The segmented flexible bow seal generates an internal wave pattern in the cushion that affects craft trim at different speeds in addition to the waveform generated by the side hulls, so optimization of these two elements has been a focus during the design development. The cushion itself can also assist with beaching and retraction operations, providing improved performance compared with a standard LCU. Durability of water jets in an environment laden with sand particles is an area where the operational trials will give helpful data.

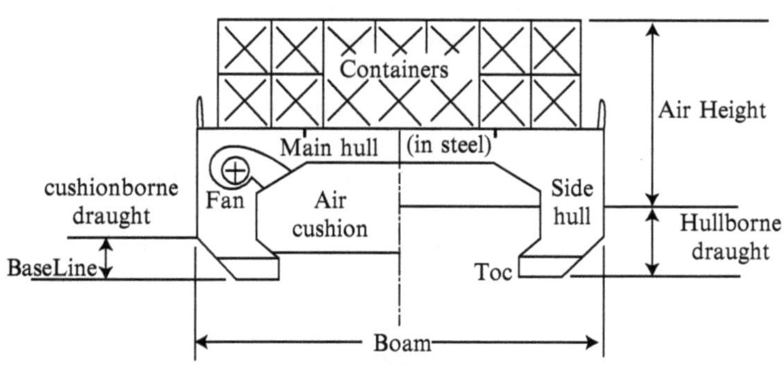

Fig. 7.18 Partial air cushion-supported catamaran (PACSCAT) X section

Fig. 7.19 PACSCAT prototype

Following design development, Qinetiq Ltd were contracted by the UK Ministry of Defence to construct a full-scale prototype Fast Landing Craft using the technology and carry out proving trials. The craft has been constrained within the footprint of an LCU Mk 10 so that operations can interface with existing Royal Navy mother ships. Payload is the same also, between 33 and 72 t depending on the equipment carried. The key difference is the demand for speeds, if possible, above 25 knots in Sea state 4, and RORO ability for the cargo for the well deck to a beach with up to 2-m surf.

The prototype shown in Figs. 7.19 and 7.20 finished its initial trials programme by the project managers Qinetiq mid-2010, followed by an operational test programme by the Royal Marines in its assault landing craft role through 2011 [7-11]. It has met the design envelope requirements, and so has potential to double the rate of equipment delivery ashore from the same offshore distance, or maintain the rate from double the distance allowing effective over the horizon assault, a key target for NATO forces at the beginning of the twenty-first century. Fig 7.21 shows an artist impression of PACSCAT delivering cargo from a large Navy Amphibious Assault Ship.

IMAA Ltd, originators of the PACSCAT concept, have developed large-scale concepts aimed at container feeder services along rivers such as the Danube [7-12]. For these type of applications, the air cushion is simply a means to reduce propulsion power at a given speed. The investment in a catamaran hull form and seals will have to demonstrate a gain with reduced overall energy usage and emissions to attract a clientele. Success with the landing craft concept will be a stepping stone towards that goal.

Fig. 7.20 PACSCAT in slings

Fig. 7.21 PACSCAT with Albion littoral combat ship (LCS)

In late 2011, Damen Schelde Naval Shipbuilding in Holland began development of an air-supported landing craft aimed at faster delivery to shore from LPD (Landing Platform Dock) vessels. Their 25-m, 35 t payload craft is targeted at full load service speed of 27–31 knots compared to a conventional craft operating at about 23 knots. Their design is based on the ESI air support technology (see earlier in this chapter).

M Craft

M Craft can be traced back to a small boat in Italy that had been running in the Venetian Lagoon aimed at reducing the wake and the destruction to the canal banks made by the traditional boats' wash. The founders of M Ship Co in the USA realized that if the wave pattern from the hull could be cancelled on the outside and suppressed by a reducing tunnel from inside, resistance should be lower, and developed a novel craft with bottom profile shaped like the letter M.

M Ship Co in the USA have developed several single M planing tunnel craft and had great success in trials, demonstrating that the craft performance was better than that of the Planing Tunnel craft (PTC) described in Chap. 6, sometimes called the W Craft, since the bottom profile is like the letter W. Following successful tests of the single M craft, M Ship Co developed the double M craft, shown in Fig. 7.25 [7-13].

M Craft Configuration

Figure 7.22 shows typical lines of a single M craft, and Fig. 7.23 shows a Double M Craft, the 27-m "Stiletto", built in 2006. Figure 7.24 shows a cross-section of the single M Stiletto hull. In the figure can be seen the central hull which is configured for planing and the thin side hulls. The tunnels inside the side hulls are of conical form. The internal geometry of the planing tunnel is formed as a spiral configuration so as to capture and reflect the energy generated by the bow wave and spray to increase the dynamic lift and decrease the wetted surface area, and so decrease the resistance and improve the hydrodynamic performance, see Fig. 7.25. The external configuration of the hull is a series of almost flat surfaces in vertical and longitudinal planes. The leading particulars of double M M80 Stiletto are:

Loa		27 m
Boa		12 m
Height		5.6 m
Draught	Static	0.8 m
	Dynamic	0.457 m (at 50 kt)
Displacement		67.1 t
Main engine		4×Caterpillar c-32 diesel with 1,213 kw each
Speed		50–55 kt in calm water, 35 kt in SS 3–4
Payload		20 t
Personnel		15 (including crew 3)
Hull structure		FRP

Fig. 7.22 Single M craft

Fig. 7.23 Double M craft

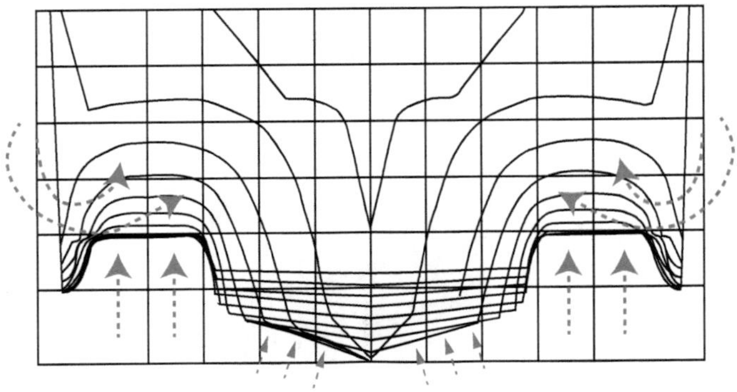

Fig. 7.24 Stiletto hull cross-section

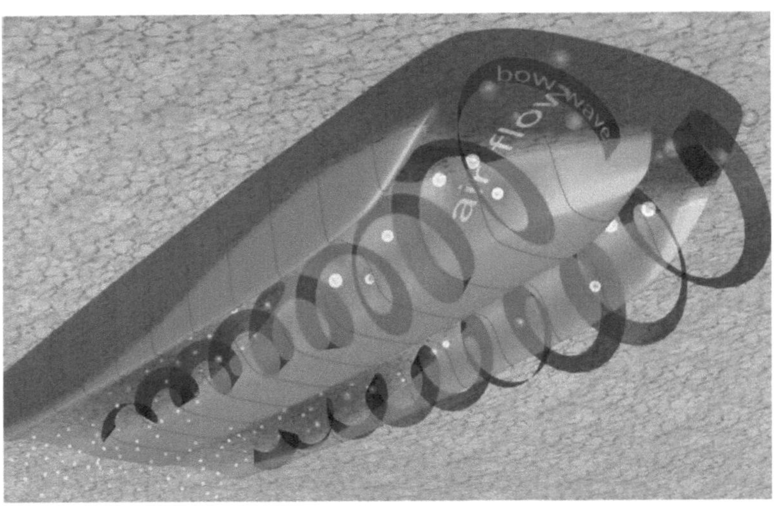

Fig. 7.25 M craft spiral between hulls

Hydrodynamic Efficiency

Since the bow wave and the spray caused at high speed will be captured and reflected by the hull internal form, the wake and spray will be greatly reduced. The physical phenomena can be explained as the bow wave and spray are pressed under the tunnel and into the tunnel longitudinally as an outward rotating spiral relative to vessel centre line, thus two phase flow (both water and air) is generated so increasing the lift, and by using the wave and spray energy to lift the craft, thus decreasing the wetted surface and both wave and friction drag.

Fig. 7.26 M craft wake vs. RIB boat

Fig. 7.27 M craft Stiletto at speed

Figure 7.26 shows the comparison of the wake of the Stiletto in the background with a conventional planing craft. The Stiletto is propelled by two Arnesen Surface drives similar to those shown in Fig. 4.13. The Arnesen drive support structures and indeed the stirred wake are evident in Fig. 7.27.

Seakeeping

Due to the lift acting on the tunnels generated by mixed flow of air and water vapour/spray, the damping force to motion in waves is improved and so impact forces in waves are decreased. In addition, the transverse restoring force in rolling motion is increased due to the wide distribution of air water mixture flow, thus decreasing roll angle and improving transverse stability.

The M craft should have a very low underwater noise characteristic and reduced induced wave making due to the decrease of wake and spray caused by the craft. In addition the reduced wave making and wake pattern will cause less disturbance at speed to other traffic in confined waterways. Meanwhile, the main disadvantage of the M craft is high cost due to the rather complicated structural configuration.

Future Prospects

We have reviewed a mixed bag of concepts in this chapter. Combining the different lifting forces has lead to some very successful craft—the HSS being a striking example commercially and the Sea fighter among military craft. The more sheltered marine environment in Russia has inspired another interesting vessel in the FACAT as well as the ACC, which has been a successful competitor to the shallow submerged hydrofoil craft. In the West, the ACC is only now showing some promise, though it may still remain a niche compared to catamarans. One really interesting development over the last decade has been the use of stabilizer T foils on all sorts of catamarans, particularly the wavepiercers—the FACAT is a kind of extreme example of this where the foils are used to lift the craft bow right out of the water. Even without going that far, it is clear that adjustments to the fore body design of a catamaran and adding stabilization is a useful approach to optimization of these basic vessels.

What about the air cushion? While the ACC clearly can give improved performance on lower power in the right conditions, the environment they are suited to is a less challenging one—rivers, estuaries and coastal conditions for high-speed operation anyway. The deeper cushion of the PACSCAT may be helpful for the assault landing craft mission, but the complicated structure, combined with the air cushion machinery, don't sound too attractive for longer range missions compared with the alternative of a high L/B catamaran with simpler lines, or the modification of a monohull with a shallow cushion cavity under a reconfigured shallow double bottom of the hull. It will be interesting to see how PACSCAT develops as it transitions from prototype to operational vessel.

The air cushion-supported hull form developed by Effect Ships International appears to have significant promise with its power saving compared to equivalent planing monohull and is beginning to be taken seriously by the leisure craft market in 2011, and once its seakeeping comfort is also appreciated compared to the harder ride of the normal craft, this could create a breakthrough. The existence of bespoke craft such as the Millenium 140 is also an encouragement that the technology can be taken further.

Strangely, the M craft may literally be the dark horse of the pack. While the hull form is somewhat complex, at the end of the day it is simply a structure. The platform for the craft is close to rectangular which is fine for outfit arrangements, and the propulsion, whether water jets or surface riding propellers as used for Stiletto, is straightforward. The potential high speed and high payload should be attractive, leaving only the potential to scale the hull constructed in fibre/resin as its limiting factor, together with the required environmental conditions to achieve the particular mission.

Returning to the catamaran SWATH concepts, while the HSS was very successful, and the X craft also in fulfilling its role, their advantage is small compared to "normal" large catamarans now that the semi-planing geometry has been refined by designers, so overall it is simpler for an operator to stay with a "standard" design. This can also be important when considering the longer term and the fact that most fast ferries are passed on to other operators after a number of years and a fast ferry route matures attracting more business.

We are now beginning to discuss the subject of our next chapter, the HPMV Market, so should really move on! The message around these hybrid craft seems to be that they are most successful when responding to a niche market or special mission. Let's turn to the HPMV Market as a whole and how our complete collection of concepts fits in.

Chapter 8
HPMV Market and Future

Previous chapters have looked in some detail at the different types of high-speed vessel that have developed to date and the markets that have driven them so far. Their development has not been independent, as we have seen. The Hydrofoil, Hovercraft, and sidewall hovercraft (SES) types have developed in the same time period and have competed with each other on river and coastal passenger ferry services. Apart from ferry services, the different HPMV types have tended to mature into vessels that fit a particular niche application, whether that is passenger carrying, a utility task, or a particular military mission. In many cases, the application has been a targeted improvement on an existing mission envelope. A number of programmes, mostly those sponsored by governments through military or public requirements, have tested the performance limits of craft through prototypes development programmes, for example the early hydrofoils and hovercraft and more recently catamarans and trimarans, and then edged back the specifications for the operational procurements.

When selecting a type of HPMV for a particular mission, a primary requirement is the best combination of economic efficiency and attractiveness to the customer, whether this is a passenger transport, or a military mission demand. Let us look at how the currently achievable attributes of different HPMV compare to give a little more understanding for the market dynamic, and perhaps give us a direction for the future.

HPMV Capability Contrasted

As a starting point, we present Table 8.1 on page 288 for readers' reference, giving qualitative assessments of the main characteristics based on current technology. In the table, we use five grades for comparison, i.e. "excellent, fine, good, fair, and low", for available payload volume we use high, medium, and low to distinguish between them.

The table is the authors' assessment based on current development of the different craft. Some characteristics are generic, such as payload volume which is controlled by the geometry of the craft type. Other characteristics are able to be changed such as

L. Yun and A. Bliault, *High Performance Marine Vessels*,
DOI 10.1007/978-1-4614-0869-7_8, © Springer Science+Business Media, LLC 2012

Table 8.1 Comparison of key characteristics for various HPMV

	ACV	SES	Monohull	Hydrofoil	CAT[a]	Trimaran	SWATH	WIG
Speed	Fine	Fine	Good	Good	Fair	Fair	Low	Excellent
Fuel consumption	Fair	Fine	Good	Fine	Good	Good	Fair	Fine
Seaworthiness	Fair	Good	Good	Excellent	Good	Fine	Excellent	Excellent after take-off, poor before take-off
Working and passenger space	Fine	Fine	Low	Low	Excellent	Fine	Fine	Low
Amphibious capability	Yes	No	No	No	No	No	No	Yes, DAC WIG, DACC no for others
Take-off sensitivity to weight	Very sensitive	Sensitive	Sensitive for planning hull, less sensitive for round bilge craft	Sensitive	Sensitive	Sensitive	Sensitive For CG	Very sensitive
Cost	Fair	Fair	Excellent	High for DSACHYC	Excellent	Fair	Fair	High
Payload percentage	Fair	Fair	Excellent	Good	Good	Fair	Fair	Low
Payload volume	High	High	Low	Low	High	High	Medium	Low
Technology complication	Fair	Good	Excellent	Fair for DSACHYC	Excellent	Fine	Good	Fair
Maintenance	Fair	Good	Excellent	Fair for DSACHYC	Excellent	Fine	Good	Fair

[a]Including wave piercing catamarans

the technology complication as we call it, so that a craft may become more attractive. We have discussed these in earlier chapters.

A weighted analysis of different characteristics will need to be applied during craft type selection by a designer or potential operator, based upon analysis of the importance of the characteristic to the mission. On most commercial missions, the vessel speed in average environmental conditions of wind and sea state are important, so as to be able to maintain a reliable service, and thus define the economy of the operation. Directionality of the environment is also significant as cross seas and winds strongly affect seakeeping, and may force a change of the route if the vessel characteristics are not compatible [8-1].

For commercial craft, the purchase and operating costs, cabin space area/volume, and maintenance complexity, will normally be controlling factors since the potential operator's budget, and current terminal facilities will guide what is most practical to introduce in to service. The timeline for introduction is often an important factor, so as to allow trials prior to high season operations. The purchase will most likely be funded by a bank loan and the ability to repay the loan in the early years of craft operation will be important to obtain the loan in the first place.

In the case of military craft, first cost and operating cost per operating hour are not so critical and can allow higher aspirations for performance and lower for endurance at high speed, while amphibious capability can be important for certain applications, and efficient loiter performance or lower speed extreme endurance also often enter the selection criteria. The controlling factor will be the lifecycle cost including the maintenance and refit costs through the vehicle life.

In both cases the ability to operate successfully in a rough sea, the seaworthiness, is important. In the case of a passenger craft this relates to the number of trips that may be cancelled for bad weather while for military craft it will represent the wait on weather time that might be experienced before the troops and equipment can be landed or the transit time to forward sea base for a littoral combat or logistics vessel.

An offshore cruising yacht or super-yacht has a slightly different economic equation, since the use at high speed will be limited, as the majority of the service life will be spent at rest or slow speed operation. It may well be that when offered for charter, the price on a daily or weekly basis is controlled by the capital amortization cost due to the very high investment in the craft internal outfit to provide the luxury environment for the owner, and charter parties. The mission cycle has to be analyzed and costed in a similar way to any ferry or military mission, and appropriate margins allowed for taxation, eventual replacement, etc. in order to determine the cost per passenger mile, operating day, or guest place per day.

Marketing Analysis

HPMV operations, particularly fast ferries, have grown quickly in the last half century, as many of the world's largest cities are at the coast and have routes to population centres close by, along rivers or across straits. Traffic development has been controlled by personal wealth, which in turn has controlled people's travel habits and desires, and their ability to support ferry operation costs. We have seen the market develop and mature in the Mediterranean, Coastal Europe, Japan, Australia, China, and more recently in North and South America. Other areas of the world have seen operations briefly, often curtailed by optimistic assessments of market size, or practical ticket price. As the world economy develops, the ferry market will continue to expand. Business success for an HPMV builder or operator is therefore controlled by accurate market projection and having a good product fit rather than necessarily having a product that makes a step change.

For the HPMV designer and builder market potential is important for successful development of a craft to provide leverage on the initial development cost. The most successful designs are ones that can be replicated a number of times, and perhaps scaled up, allowing costs to be driven down from the initial prototype and performance enhanced in steps that are manageable from the investment point of view. An operator on the other hand is looking for a craft that can give reliable service for at least the period that his financing allows him to pay off the loan and set aside sufficient funds to make the next upgrade to his fleet as the market develops. This will include the sales price for the craft when it is passed on to another developing operator.

These issues are typical for any successful business development, and the reader should be able to find guidance in many business management textbooks. The key is to have accurate data to build the business model, and have a good margin both for the earning power and for the investment costs. HPMV are a high cost low volume business, and so losses on any individual vessel delivered can be catastrophic for the business, whether a builder or operator. Time and again projects using prototype or early production craft have found costs rise more than the optimism encouraged by enthusiasm originally projected. A prototype or initial production unit is after all a learning experience, and learning costs money and time. A helpful guideline is to look at the project at double the cost in all the main areas, and if it can withstand that kind of change then it should be robust. Refining down from there demands significant market numbers to distribute development cost.

Cost reduction curves have been developed—the US Navy did that for its landing craft, air cushion (LCAC) programme. The LCAC had a predictable market since the target was set up before the programme began—a fleet of around 50 craft on each of the west and east coasts of the US. For a commercial development this may be more difficult unless the builder can afford to prime his market by building "on spec" and sell the craft during the build phase. The fast catamaran builder Incat has used this business model successfully for many years. In the earliest phases of development, the R&D cost is difficult to justify and fund "internally" within the development. This is where sales of existing craft or their operation need also to provide a portion of funding for future craft or operations development for an established builder or operator.

Funding for prototypes generally comes from bank loans or from a government if the mission is military or paramilitary. In the first situation the bank will require a

Table 8.2 HPMV Newbuilds by year 1987–1994

Year	1987	1988	1989	1990	1991	1992	1993	1994
WPC	10	10	12	8	6	6	6	5
Foilcat	0	0	0	2	3	9	13	10
Catamaran	60	65	56	60	53	47	72	73
ACV	2	3	7	7	4	2	0	1
SES	11	24	14	17	10	7	11	6
Hydrofoil	17	19	27	25	24	15	11	7
SWATH	0	3	4	1	0	2	1	1
Monohull	25	21	18	22	13	17	23	20
Total	125	145	138	142	113	106	137	123

solid business plan that is based on sales of craft, or passenger/vehicle trips. This can be difficult when the HPMV is building a new market so it is where the "stress test" suggested above can be useful, and give confidence to the funding suppliers.

Another approach to funding technology developments is to use development funds available for technology research such as those in the European Union. Development of the air lubricated fast craft by Effect Ships International has been supported in this way. The hybrid WIG and Hydrofoil Ferry has also been developed as a concept under the HYDAER programme of the European Union. In Japan a number of HPMV developments have been supported through government funding and in the UK much of the early development of ACV was supported through the government. Since 2006 South Korea has been funding an industry wide development for commercial WIG craft, supporting a number of companies to move to full commerciality over a decade of development work. Such funding can be used as a lever to move a technology from experimental to practical, and is important where the required investment is large. Nevertheless, for long-term success the product has to be economic as well at technically exciting.

The HPMV market has developed relatively quickly since the 1960s. Many services and operations have been established and then found not sustainable in the long term. Equally, operators with popular routes have found the need to refresh their fleets every 5–10 years, much shorter than a vessel useful life. This liquid situation has developed a market for "pre-owned" craft that can be moved to less challenging environments supporting the upgrade path for the tougher services.

Careful analysis of the existing and potential future market for the proposed craft should therefore be completed before finalizing the design specification. This will need to include current demands for passenger, vehicle and other cargo in terms of ride, accommodation quality and "ticket price". On the operator side it will need to include current and future demands for energy efficiency.

Tables 8.2–8.4 give some statistics for the HPMV Ferry market since 1987, showing the main trends for craft type and size. This gives an indication of how the market has moved in recent years. Figure 8.1 shows a chart of craft delivered in the 1987–1994 period, while Fig. 8.2 shows more recent data over the last decade. In the earlier

Table 8.3 Deliveries of HPMV, world (except Russia) over last decade

Year	2000	2001	2002	2003	2004	2005	2006	2007	2008	2009	2010
Craft type summary											
WPC	3		1	1	1	1	1	3	1	1	1
CAT	33	31	45	37	39	29	39	45	39	34	32
Trimaran						1					
Hovercraft			1				1	1			
SES			1		1						
Hydrofoil	1		4	1		4	3	1	1		
Monohull	5	12	9	10	2	4	4	7	3	2	14
SWATH			1								
Total	42	43	62	49	43	19	48	57	44	37	47
Passenger ferries											
50–99 Seats	1	6	11	6	2	8	10	8	9	4	14
100–149	7	7	7	8	14	5	17	8	6	5	7
150–199	5	5	16	4	6	6	1	14	4	3	2
200–249	2	1	5	11	3	7	4	8	3	7	5
250–299	0	4	6	3	2	0	4	1	2	4	6
300–349	6	2	1	1	2	5	5	2	0	4	1
350–399	3	6	3	2	2	0	1	1	1	2	1
400–449	4	5	9	3	2	1	1	8	10	4	7
450+	0	1	1	6	4	1	1	0	1	0	1
Total	28	37	59	44	37	33	44	50	36	33	44
Passenger/vehicle ferries											
5–49 cars	2	1	2	1	2	3	2	2	3	0	2
50–99	3	0	0	3	3	1	1	0	2	2	0
100–149	0	0	0	0	0	0	0	0	0	0	0
150–199	2	2	0	0	0	1	1	0	2	0	1
200–249	2	2	0	0	1	0	0	3	0	0	0
250–299	4	0	2	1	0	0	0	1	0	1	0
300–349	1	0	0	0	0	1	0	0	0	0	0
350–399	0	0	0	0	0	0	0	1	1	0	0
400–449	0	1	0	0	0	0	0	0	0	1	0
450+	0	0	0	0	0	0	0	0	0	0	0
Total	14	6	4	5	6	6	4	7	8	4	3

period, the number of craft delivered was much higher than this last decade, and the spread of craft types was greater. More recently, the focus has been on catamarans, and a steady market for the mid range passenger craft. This is illustrated in Figs. 8.3 and 8.4 showing the deliveries against passenger and vehicle capacity.

In the early 1990s the oil price was stable and began to diminish, encouraging a significant growth in personal travel. Through the last decade, fuel costs have risen rapidly, with very high spikes since 2007, so HPMV operators have been in challenging times for operating cost.

The key craft type for commercial service, both passenger and mixed passenger and vehicles is the catamaran, which has been able to follow the market by scaling

Table 8.4 High-speed ferries in operation in top ten countries (excluding Russia)

Countries (districts)	TRI	CAT	HYF	Monohull	SES	ACV	WPC	Total
1993[a]								
China mainland		73	22	13	21	14	0	143
Hong Kong		54	18	9	37	0	0	118
Italy		6	90	9	1	0	0	106
Japan		22	31	38	0	5	1	97
Norway		47	0	16	4	0	0	67
Greece		1	47	0	2	0	0	50
USA		19	0	18	0	0	0	37
Australia		18	0	8	0	0	4	30
UK		4	2	0	1	8	6	21
South Korea		4	5	0	12	0	0	21
Other countries		61	50	31	6	5	0	153
Total		309	265	142	84	32	11	843
Countries (districts)	TRI	CAT	HYF	Monohull	SES	ACV	WPC	Total
2010[a]								
China mainland	0	103	24	33	21	14	4	199
Hong Kong	0	84	21	11	30	0	0	146
Italy	0	16	95	12	1	0	0	124
Japan	0	32	31	41	1	2	2	109
Norway	0	50	0	18	0	0	0	68
Greece	0	11	50	0	2	0	0	63
USA	0	24	0	30	0	0	1	55
Australia	0	28	0	12	0	0	6	46
UK	0	8	2	0	0	10	6	26
South Korea	0	12	5	0	12	0	0	29
Other Countries	1	344	52	57	8	7	6	475
Total	1	712	280	214	75	33	25	1,340

[a]Authors' estimate

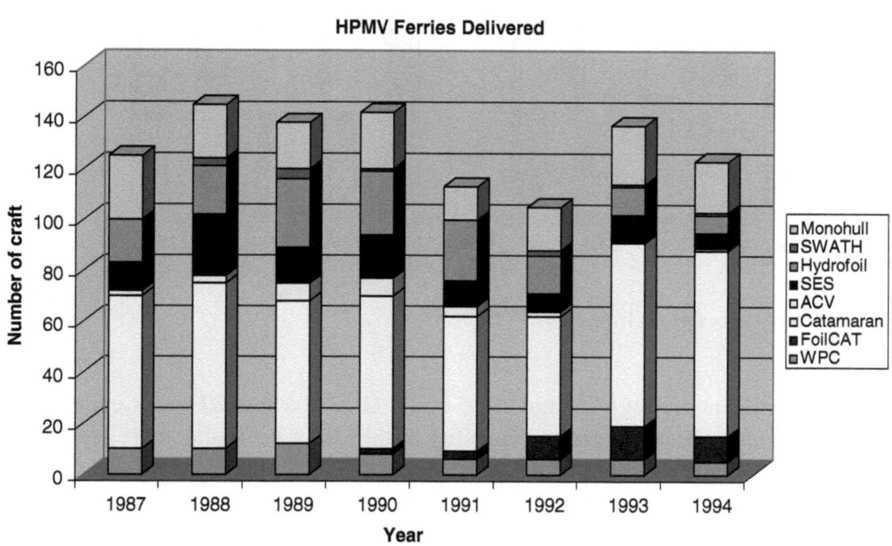

Fig. 8.1 Fast ferries delivered 1987–1994

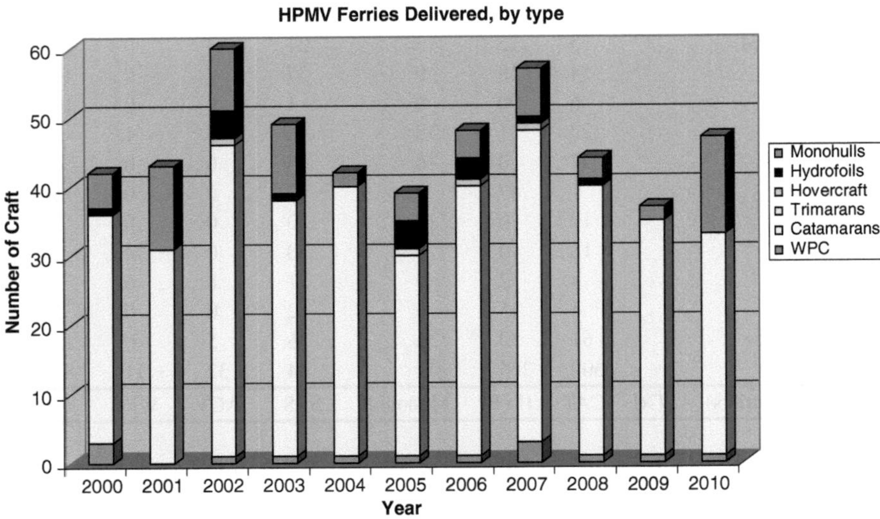

Fig. 8.2 Fast ferries delivered 2000–2010

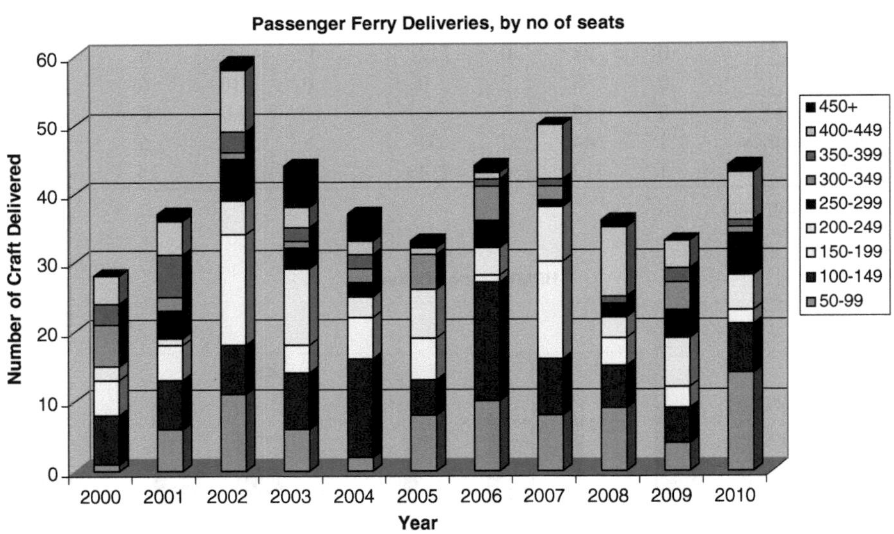

Fig. 8.3 Passenger ferry deliveries

upwards with increasingly competitive costs. Over the last 20 years catamarans have developed ocean traversing capability as they have been scaled up. They have demonstrated their ability to traverse the world's oceans, mainly on delivery trips, and have taken the Atlantic Blue Riband Trophy.

The Ferry data presented above is one segment of the HPMV market only. It is important as one of the drivers for the HPMV industry. The other major drivers

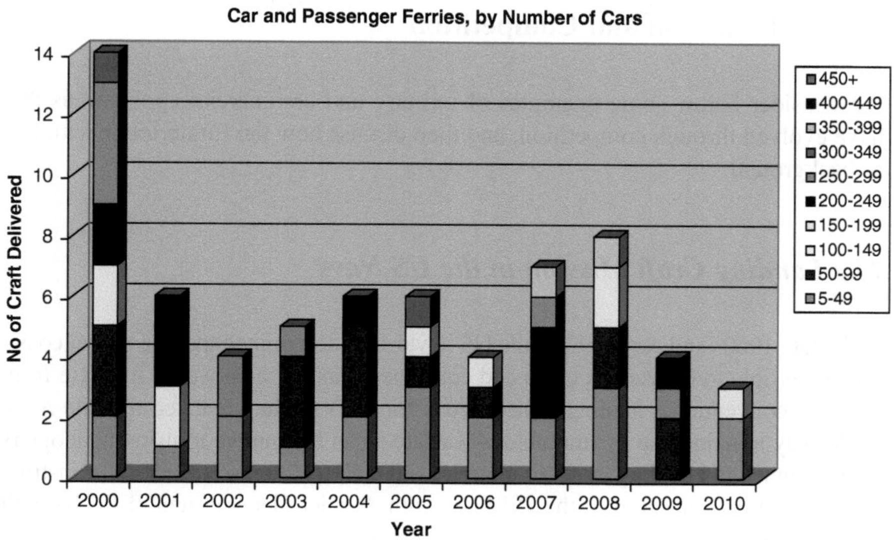

Fig. 8.4 Passenger/car ferry deliveries

are the military market, and the utility market. The military market has been an important support to development of hydrofoils, ACV and WIG craft as we have described in previous chapters. In the last decade the focus for large vessels has been on catamarans and trimarans, based mainly on US armed forces requirements. The utility market has been the area that ACV's have become accepted for a number of specialist roles and this has resulted in a steady stream of orders. The monohull world continues to provide steady business for small and medium shipbuilders. Not so many fast ferries as can be seen from the figures above, while an increasing number of small shipbuilders have broken in to the "super-yacht" market by teaming with specialist Naval Architects and interior design houses.

The British Navy has tested a prototype trimaran warship and the US Navy has tested both the trimaran and catamaran over several years followed by commitment to multi-vessel programmes for both the Littoral Combat Ship and Joint High-Speed Vessel programmes that we discuss later in this chapter. The catamaran is now entering service with the US Navy for fast logistics in the JHSV programme and the trimaran in the LCS programme after almost a decade of evaluation with leased craft. Military development takes much longer than the commercial market. We discuss this most recent development later on in this chapter in comparison with some of the earlier programmes with hydrofoils and ACV's.

We have seen in earlier chapters how Britain, the USA, and the USSR have invested large sums in developing HPMV technologies at smaller scale, hoping to be able to scale up, but not being able to reach that goal in most cases. More often, it has been the niche application that has been successful. In a sense, the LCAC is just that—a very specific mission with a unique solution that is now actually quite old, not far from the same vintage as the Boeing 747 aircraft, so let us review the LCAC first.

HPMV Evolution and Competition

We consider below some examples of military and commercial applications that have evolved through competition, and then discuss how the future looks with this as background.

The Landing Craft Mission in the US Navy

In the late 1960s and early 1970s, the US naval staff determined that the global coastline accessible to be used as troop and equipment landing area would increase from 17% for conventional landing craft to 70% for ACVs. They realized that the ACV could play a major role in amphibious warfare as an alternative to using helicopters for personnel and landing craft for materiel and tank transfer, so decided to construct two competitive prototypes, the JEFF(A) with Aerojet General, and JEFF(B) with Bell Aerosystems, see Fig. 2.15. After a series of trials including tropical and arctic environments they decided to use the prototype craft JEFF(B) as the basis for the amphibious landing craft series LCAC. This prototype has since been developed up to operational specification with 100 craft built through the 1980s and since that time deployed as fully operational units with the US marines. LCAC's have been used in many parts of the world since the early 1990s, including Somalia, Kuwait, and the Far East. The craft are based with larger ships—the LHD class—and deliver their cargo to shore from these, stationed just over the horizon from shore. They participate in annual exercises and often if there is an environmental catastrophe such as a subsea earthquake followed by a tsunami, the LCAC's are used for emergency logistics. Figure 8.5

Fig. 8.5 Landing craft, air cushion (LCAC) operation in Sumatra

shows logistics support in Sumatra following the Tsunami from the Indian Ocean earthquake at 0.58 on December 26th 2004. The waves were many metres high as they swept into coastal areas like this and devastated the coastal villages. The LCAC were used to take in emergency supplies to a population in shock and without means to support themselves in the immediate aftermath.

The LCAC was able to be developed to fit the mission requirement within the Navy budget and operating cost requirement partly through the market size of more than 100 vehicles. Continuous improvement processes within the manufacturer and the US navy operations support organizations has enabled cost savings assisting its evolution over time to become a core asset of the US Navy. The cushion system has been improved through an enhancement programme late in the 1990s, and since then the Navy has started to look at its successor, either an enhanced specification for a direct replacement, or an alternative.

Since 2007, the Office of Naval Research has run a programme for a much larger craft which would fit with a target to deliver "development of significantly enhanced enabling and long term capabilities for the Joint Sea Base and Ship-to-Objective manoeuvre".

While for the established missions the ACV development can be by stepwise improvement of the components, the US Office of Naval Research programme is a more radical development to make a step change in its "Sea Base" capability. This involves bringing together elements of current catamaran and SES technology and a new approach to amphibious cushion design so as to create a "Transformable Craft" that can deliver a heavy payload directly ashore from over the horizon to unprepared terrain [8-2].

The US Navy already has amphibious capability with the LCAC, but the payload capacity of the LCAC cannot deliver the desired hardware ashore within a reasonable time from over the horizon—craft with capacity of 4–10 tanks. The LCAC has a payload capacity of approximately 70 t, while the US Navy objectives are a threshold of 300 t and ultimate objective of 750 t—in the same range as the JHSV payload. In contrast to the JHSV though, the payload will be taken on board close to the objective and shuttled ashore. Clearly a craft of this size will be much larger than an LCAC which is already the limit for current well deck ships in the US Navy and so would have to travel alongside the mother ship to the "Sea Base" before use as the shuttle ashore.

The idea being researched is a craft that can operate as a catamaran or partially air cushion supported craft for transit from home base up to 2,400 nautical miles away (or more by way of refuelling) at 20 knots or so, and then to receive the payloads by ramp from a well deck, or crane by side loading, transit over the horizon to 1 km or so offshore as an SES at approximately 40 knots, and finally deploy a full peripheral skirt system, and changing from water jet propulsion to air propulsion transit ashore where the payload is discharged, see Fig. 8.6. There are three large industrial consortia working in competition to develop prototypes which will be tested, before a procurement programme is started, led by Textron, UMOE America, and Alion Industries, respectively.

It is interesting to compare this challenge with the PACSCAT programme under way in the UK, which targets payloads in the LCAC range and delivery speeds

Fig. 8.6 ONR Transformable Craft—Textron Beach scenario

rather lower than LCAC while in similar sea states. The PACSCAT is an evolutionary concept—trying to extend capability while maintaining the base concept that the UK Royal Navy can absorb, rather than creating a discontinuity.

The T-Craft, on the other hand, is a "game changer" concept. If it works out well, the US Navy will have a completely new tool to work with, rather like it achieved with the LCAC in the 1980s, and will enable some step changes in operational capability to be achieved.

Technical focus areas for all the consortia developing this new technology are high power density air cushion fans and water jet propulsors, the transformable cushion seal systems, and high power electrical prime movers so as to be able to switch between water jets and air propulsion. Practical systems for the offshore payload transfer, and innovative structural design are also an important part of the development. A test model of the Textron concept coming ashore is shown in Fig. 8.7.

Major technology steps such as this are rarely seen in the commercial world, partly due to the considerable risk involved, both technical and commercial. The consortia participating in the programme will all gain significant knowledge and experience which may then be applied for utility or commercial application. The T craft faces technical challenges at all stages of its deployment mission from home base, and an extreme case is the final delivery stage where military resilience is a strong driver for the design, [8-3]. In this relation, the hull structure may well need to incorporate elements that are both metallic and resin/fibre composite. Much military armouring is now composite material, while large scale structures for fast craft are generally in aluminium for craft of this size driven by cost considerations.

Fig. 8.7 ONR Transformable Craft—Textron model testing

The T craft programme was started in 2007 with concept studies by all competitors continuing through the technology development phase II which includes detailed development of the cargo loading interface offshore to the transport ships, and full scale mid section structures for testing. From 2011 through 2014, a final concept is expected to be selected followed by prototype demonstrator construction and operational trials. If the JHSV approach is followed, there will also need to be a second supplier for the deployment programme.

Ferry Routes as a Driver for HPMV Development

We start with a look back at ferry services on the Channel between England and France. We then look at the Pearl River delta between Hong Kong, Macau, and Guangzhou and finally at the Taiwan Strait before discussing ferry development in other parts of the world.

English Channel Ferries

The British Hovercraft SR.N4 passenger/vehicle hovercraft commenced trials in February 1968, and made its first channel crossing from England to France on June 11 that year. The SR.N4 was the first truly open-water passenger/car ACV ferry craft capable of all year round service over an open sea route where waves of 2–3 m can be encountered. It has achieved speeds in excess of 90 knots over calm seas and

Fig. 8.8 SR.N4 cross channel operation

operated out of specially designed terminals at Ramsgate, Dover Boulogne and Calais delivering passengers and cars across the channel faster than services though the Channel tunnel that was opened 25 years after the craft first entered service. Figure 8.8 shows an SR.N4 in service on the English Channel.

The Hoverlloyd and British Rail Seaspeed services set up at that time were truly game changing compared with the existing traditional ferries. During the main summer holiday season, with good weather they were able to deliver a really fast service. In winter time though, the heavy sea states meant that at times services had to be cancelled. Traditional ferry services were also cancelled, so it was not a special problem to the hovercraft, just accentuated. For a while it appeared the hovercraft would begin a trend that would be replicated more widely in Europe and elsewhere. The aerospace technology used in SR.N4 meant that normal ferry operators were not willing to make the commitment in terms of maintenance, and indeed in building completely new terminal facilities, and so this did not happen.

At the end of the 1990s, car–passenger wave piercing catamarans began operations in English Channel and competed with SRN4 services. While slower than the hovercraft services, the catamarans were able to continue services in weather that the hovercraft had to cancel, and so the reliability became a selling point. This was strange in some ways since most people cross the channel during summer for their holidays when the weather is fine and the channel can often be glassy calm! Nevertheless, the catamarans' internal outfitting was more up to date, with lower noise in the cabin, more space to walk around, and for the operator lower maintenance costs. It was not a simple win though—the initial operations showed that the catamaran structure was being worked to its limit during the winter season and so repairs were necessary to improve the structural design. The operating costs

Fig. 8.9 WPC cross channel operation

based on marine diesels became the deciding factor, as the gas turbine powered hovercraft eventually were not able to maintain economic competition as fuel prices steadily increased.

The inauguration of the channel tunnel also changed the competitive landscape on the channel, as all ferry operators had a new business to compete with that simply had to take a large volume of business in order to pay off the huge investment cost. Large catamarans are still able to compete for this market, particularly for clients living locally to the coast, as the channel tunnel is oriented to take traffic from the motorways at both entrances in England and France. This story demonstrates once again that in the long term a simple configuration, easy to handle and maintain, using marine engines and equipment as well as marine construction, are most important factors for long-term competition and marketing. Figure 8.9 shows a 98-m Incat car–passenger WPC "Normandie Express" that operates on the English Channel.

Pearl River Delta Ferries

There are many types of HPMV currently operating and competing on routes around Hong Kong/Macau and Pearl River Delta, as shown in Figs. 8.10–8.13. Figure 8.10 shows the Wavemaster 39 m CAT, Fig. 8.11 the Wavemaster 42 m CAT, Fig. 8.12 the Fjellstrand FoilCat 35 m HYCAT, Fig. 8.13 the CSSC PS 30 Hydrofoil, and Fig. 8.14 the Jetfoil 929-100.

This ferry competition for high-speed craft took off in the 1960s partly due to Macau developing as a leisure destination for Hong Kong people at the weekend,

Fig. 8.10 Wavemaster 39 m

Fig. 8.11 Wavemaster 42 m

and for gambling. As the political situation improved in the 1980s and 1990s, traffic to Shenzhen and Guangzhou steadily built up. There are many millions of people living in this area and so demand is rather high.

Hovermarine Ltd in the UK built the first successful SES ferries at the end of the 1960s and had great success with passenger ferry operators between Macau and Hong Kong due to its combination of low power and high speed relative to the traditional ferries operating at that time, giving attractive route possibilities to the

Fig. 8.12 Fjellstrand 35 m Foilcat "Barca"

Fig. 8.13 CSSC 30 m Hydrofoil PS30

traditional ferry operators. Eventually 19 HM-2 SES were operated by the Hong Kong and Yaumati Ferry Company on various Hong Kong routes at the peak of use of this craft. Figure 8.15 below shows the HM-2 fleet running in formation in Hong Kong harbour.

Fig. 8.14 Jetfoil 929 Turbojet Urzela

Fig. 8.15 Hovermarine HM2 operation in Hong Kong

From Tables 8.2 and 8.3, one can see that during the 1980s there was a boom in production of SES ferries, however this faded due to limited seaway performance of first generation craft such as the HM2. Passengers used to complain about the high vertical acceleration, and so called "Bang Bang Jump" ride, even in the case of small wave height. This was due to the "wave pumping effect" (as explained in Chap. 2), which was a problem for early SES with shallow cushion chambers between the hulls, and the use of very thin side hulls. It was not until ride control systems were introduced in the 1990s that SES seaway performance became competitive with the slower catamarans. SES were gradually replaced by catamarans in the 1990s as water jet propulsion advanced in efficiency and enabled them to achieve higher service speeds while retaining lower maintenance cost. Development of wave piercing and fine bow entry hull geometries also gave a smooth ride in waves with small and medium height. The lack of a lift system enabled catamarans to have lower internal noise also, an attraction for passengers.

The route between the mainland China and Hong Kong has been booming since 1997, due to the high population at this delta area and busy commercial trade located there, and this has added to the considerable traffic between Hong Kong and Macau. Almost one third of the world inventory of fast ferry craft operate in this region, see Table 8.4. As China continues to develop, it is clear that this area will remain a core market for passenger fast ferries.

The Taiwan Strait: A Developing Prospect

In the Taiwan Strait following the gradual improvement of both politics and economy on both sides of Taiwan Strait, it is anticipated the passenger/vehicle ferry craft market will boom in next 10–20 years once relations between China and Taiwan are fully normalized. The potential demand is huge while being politically constrained due to relations between China and Taiwan for several decades. In contrast to the Hong Kong passenger market anchored with commuters and weekend tourism, traffic between the mainland China and Taiwan may develop more like the British Channel services where cars are an important part of the payload. The particular advantage with the passenger/vehicle ferry is that passengers will be use their own vehicle on Taiwan island for shopping and visiting friends and relatives as well as sightseeing, so it is expected high-speed water transportation will develop rapidly once the agreements have been established.

The distance between the corresponding cities of mainland China (Fu Zhou and Xia Ming) and Taiwan (Taipei and Gao Xiong) are relatively short, about 100–130 nautical miles, about 2.5–3.0 hours journey time at 35–45 knots, and suitable for the operation of high-speed ferries with good seakeeping. Some shipping companies, research institutes and shipyards are already beginning to prepare for such business, with proposals for a number of different ship types, such as HSWPC (using the INCAT WPC design bought by the local operator); a high-speed catamaran design project by MARIC, as shown in Fig. 8.16 [8-4]; and a SWATH design project by CSSRC as shown in Fig. 8.17 [8-5].

Proposals for services with HPMV across the strait have been developed and presented at the international HPMV conference in April, 2009, held in Shanghai, China. The operators currently delivering services in Hong Kong and the estuary of

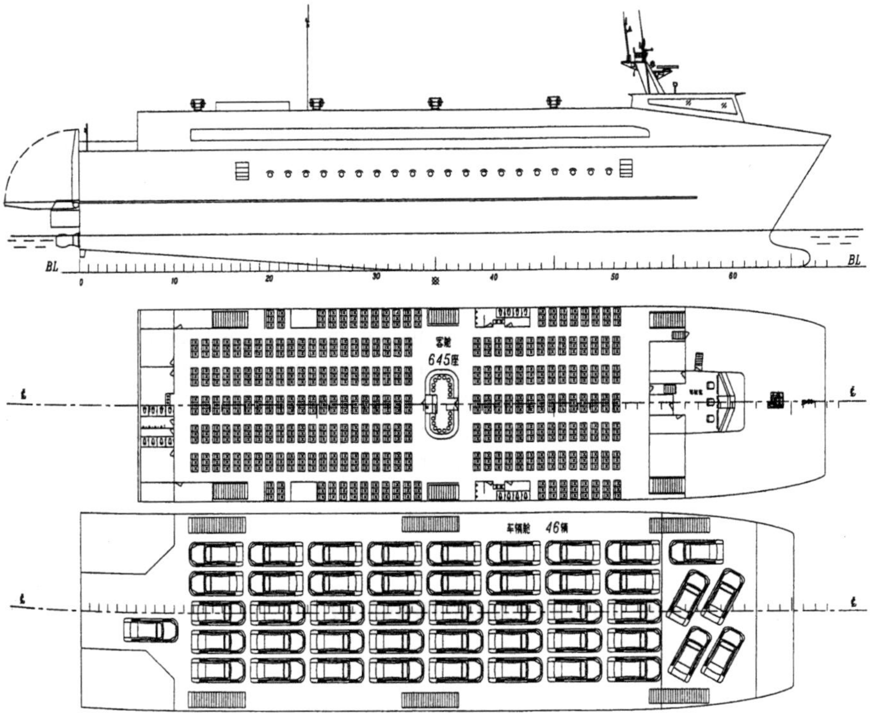

Fig. 8.16 MARIC high-speed catamaran

the Pearl river up to Guangzhou are likely to be in the forefront of developing such routes bearing in mind their local experience of HPMV built up over the last half century. The big development will certainly be the vehicle traffic on these new routes, leading to use of large size catamarans and WPC.

High-speed ferry fleet replacements outside the Mediterranean are now almost exclusively catamarans and large planing monohulls. In countries such as Norway where resin reinforced fibre construction technology has matured this is increasingly used for smaller craft. The focus is now on hull form optimisation with wave piercing forefoot designs and water jet integration in the aft part for optimized hydrodynamics. Larger catamaran craft now often use stabilizing fins both forward and aft, while monohull craft employ interrupters at the stern as a simple trimming device.

A number of accidents with fast ferries in the past 20 years have guided the industry to look much more carefully at passenger restraint in their seats, rather like the introduction of seat belts on busses, aimed particularly at journeys in rougher conditions. This attention to safety features is likely to continue in the future as the market develops further.

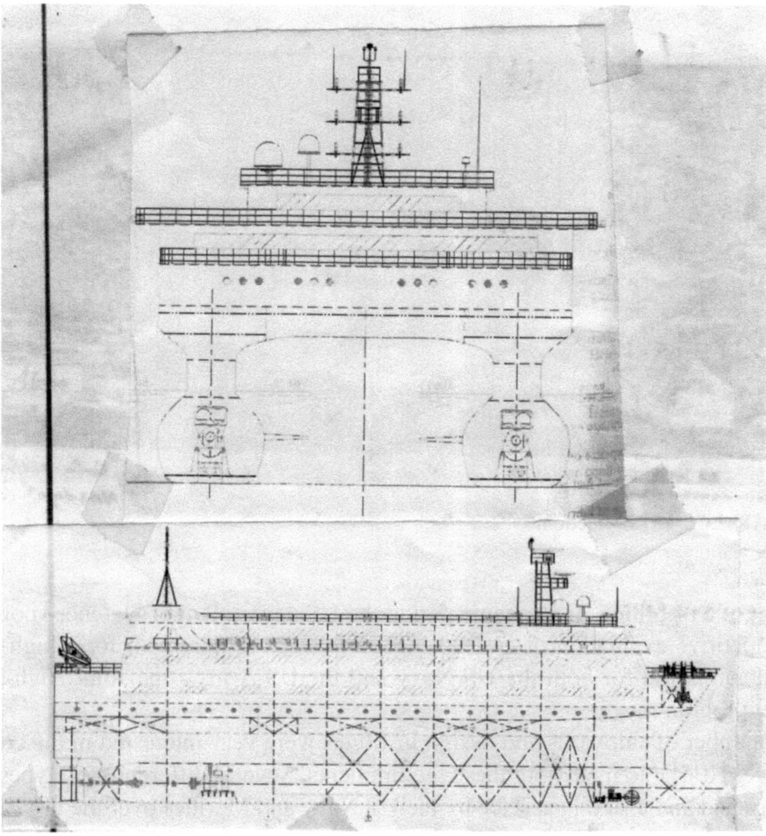

Fig. 8.17 CSSRC SWATH

Naval High-Speed Vessel Programmes

There have been a whole series of naval development programmes for HPMV from the 1960s onwards, most notably supported by NATO Nations, the US Navy, and the Russian Navy. The results of these programmes have included the NATO Patrol Hydrofoil (The USA, Italy, Israel), the Russian Ekranoplane series, the Norwegian Patrol and MCMH SES, and various monohull fast patrol craft. Like the LCAC the successful programmes still have a useful lifetime that is just a few decades before a replacement or improvement is needed.

Shortly after 2000 the US Navy began development for a new class of high-speed Littoral Combat Ship giving service speed up to 50 knots, with performance targets of high speed in waves, medium displacement (larger than 1,500 t), as well as extensive deck area aft for helicopter landing, and launching of small landing craft [8-6]. Their aim was to build as many as 50 craft up to 2017 within a total

Fig. 8.18 LCS-1 USS freedom

budget of $14 billion. In the same period the US department of Defence sponsored a competitive evaluation of craft to meet a mission requirement for a high-speed logistics mission for both the US Navy and the US Army—the Joint High-Speed Vessel programme.

A number of shipyards and design institutes were very interested in the creation of novel HPMV to meet both these requirements. Several different prototypes were constructed and tested in service by the US Navy and Marines over the last decade. For the LCS programme five concepts were evaluated:

1. *High-speed monohull craft*, "USS Freedom", Fig. 8.18: designed based on the monohull craft "Destriero", which successfully crossed the Atlantic at sufficiently high speed to gain the "Blue Riband" speed record. The craft was built by Lockheed Martin, with Loa 115 m, Boa 17.5 m, draught 3.9 m, max speed approx 50 knots, displacement 3,000 t, and service speed 47 knots in SS3 and 35 knots in SS 4–5.
2. *High-speed trimaran*: shown in Fig. 6.25 "Independence", LCS 2, was built for the US Navy in Austal's Mobile Shipyard, the USA to an order given in October 2005. Its main dimensions are Loa 127 m, Boa 30 m, draught 4 m, displacement 3,000 t, speed 45 knots, service sea state SS5 for aircraft launch, SS4 for water craft launch. The vessel carried out a series of trials in the period to 2009 when the decision on the series build was taken by the US Navy.
3. *Catamaran with small bow bulb*, interceptors at stern, stabilizing hydrofoil at bow, and ride control system. This is the "X" craft based on a design by BMT Nigel Gee Ltd, and named "USS Sea Fighter". The leading particulars are given in Chap. 7 and Fig. 7.17.
4. *SES*: The potential with this concept was to use the Royal Norwegian Navy SES "Skjøld", as basis for a new craft. The Norwegian vessel visited the USA and

Fig. 8.19 Skjold in Virginia with USN

carried out trials there for performance evaluation. Figure 8.19 shows the craft on its visit to the USA during 2009 for performance trials, leased by the US Navy while operated by Norwegian Navy Personnel. It is rather smaller than the other Littoral Combat ships tested as its mission is fast coastal patrol rather than power projection across oceans. For the Norwegian Navy the patrol mission is successful, providing a high-speed partner to the slower displacement patrol vessels in the fleet. The range is sufficient for the Norwegian Coastline and economic zone. The LCS mission on the other hand requires both a larger craft and transoceanic range.

5. *M Craft*: As an innovative craft design this has been considered as another alternative for LCS verification, see Fig 7.23. Use of this craft type would require a considerable upscaling compared to its current size, as it is in the same range presently as the Norwegian SES.

The JHSV evaluation used four vessels leased from Australian Shipyards [8-7]. Key attributes required for this mission were length less than 138 m, draft less than 4.6 m, able to transit the Panama Canal carrying at least 550 t of cargo over 1,200 nautical miles at 35 knots in SS3, and a range at 25 knots of around 4,700 nautical miles. Three of the vessels were Wavepiercers from Incat, and one catamaran from Austal:

1. *Wave Piercing Catamaran*: The US Department of Defence leased three different large WPC built to the design of Incat of Tasmania to evaluate the capability of such craft for the HSV mission, the HSV2 "Swift", as shown in Fig. 8.20, the TSV-1X "Spearhead", and the HSV X1 "Joint Venture". These craft have been operated by both the US Navy and the Army. The HSV2 has Loa 101 m,

Fig. 8.20 HSV-2 Incat WPC

Boa 26.5 m, draft 3.35 m, displacement 1,460 t, and service speed between 38 and 42 knots. HSV2 has seen service in Beirut, Cyprus, Iraq, East Timor, and supporting the joint task force operations after Hurricane Katrina in the Gulf of Mexico. HSV XI Joint Venture was in operation for the US Navy from 2002 to 2007 and saw service in most parts of the world before it was released. It has subsequently been purchased by the Isle of Man Steam Packet Company and went through a GBP20 million upgrade [8-8] before re-entering service in 2009 as "Manannan" on services to the Isle of Man. The upgrade increased maximum speed to 39.8 knots.

2. *High-Speed Catamaran*: The US Department of Defence procured a large catamaran built by Austal named the "Westpac" see Chap 6, and Fig. 8.21 for details of this craft. Westpac has been in service since 2002 and participated in a number of deployments for the US Marines. The Westpac Express has Loa 101 m, Boa 26.5 m, draft 4.27 m, displacement 1,500 t, service speed 33 knots.

LCS Programme Development

The final decision for the series build has been a result from long-term trials of "Independence" and "Sea Fighter", and shorter experience with "Freedom". The USS "Freedom" has had a more chequered career than the catamaran and trimaran so far due to the structure as-built being heavier than design, cost over-runs, and a number of technical corrections that had to be made, [8-9]. Shorter trials with the other craft types mentioned above have identified clearly their niche potential,

Fig. 8.21 HSV-1 Westpac

and so it is likely future mission requirements will see further competition from these types.

The Navy confirmed the Trimaran as their principal way forward with a contract to Austal USA for the second trimaran ship USS Coronado in May 2009. The US Navy continued with evaluation of bids through 2009/2010 for construction of the first full tranche of ten ships from the "prime contractor" whilst the second tranche of five ships will be built by a "second source contractor" utilizing equipment supplied through the prime contractor. Eventually it is planned for up to 55 LCS ships in service split between trimaran and monohull craft.

Following the design issues with USS Freedom in 2008–2010, Lockheed Martin worked hard to remove these issues from its second ship of the monohull series (LCS 3), the USS "Fort Worth". LCS 3 was launched in December 2010. At the end of 2010 the US Navy were convinced enough of the success of the monohull to award contracts for the first "LCS Flight 0+" ship plus engineering and procurement preparations for the first ten ships of the monohull series to Lockheed Martin as well as reconfirming the main commitment to Austal for the first ten trimarans. Total cost for delivery of ten ships from Austal is approximately \$4.4 billion, and \$4.6 billion for the monohulls from Lockheed Martin. Austal has new built dedicated production facilities for their ships, which has enabled them to achieve improved efficiency compared to a traditional shipyard, even considering the more complex trimaran structure.

The first LCS vessels commissioned will be based in San Diego on the Pacific Ocean rim. Operation of the vessels is planned around teams of two ships to allow for maintenance and port time, the active vessel being deployed for approximately 16 months, operated by a marine crew of 40, an air operations crew of 23, and 19 technical specialists for the modular mission packages (containerised packages that can be deployed on any ship of the class). Three crews will operate a 4-months rotation over the two ships.

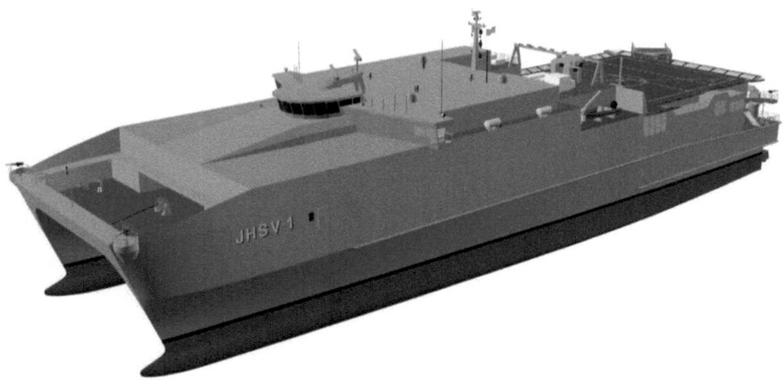

Fig. 8.22 JHSV-1 Artists impression

Fig. 8.23 JHSV-1 impression at speed

The coming decade should be an exciting one for Navy HPMV as the LCS programme matures beyond the first tranche of 2×10 ships commissioned up to 2016/2017 and moving on to the next tranche of vessels.

JHSV Programme Development

The competition between the Westpac and HSV2 mentioned above was finalized in 2009 after separate Army and Navy acquisition programmes were merged into one, led by the US Navy procurement executive (programme PMS-325). The decision for the JHSV has been a 103 m length catamaran from Austal USA, Figs. 8.22 and 8.23, with a commitment for three craft and options for another seven as described in Chap. 6 [8-10]. The operations with Westpac and the three Incat wavepiercers over 9 years has enabled the US Navy, Marines and Army to fully evaluate operating and maintenance commitments, and even take the decision to plan for commercial crewing of some of the vessels where the mission is cargo deployment to low risk destinations.

The prime contractor selected for both vessel programmes has been Austal USA. Austal has worked in partnership with General Dynamics, BAE Systems, L3 Communications, and MAPC for the LCS programme. Austal has specialized in developing efficient manufacturing techniques at its base in Fremantle, Western Australia, and this approach was extended to its facility in Mobile. The target of both supplier and customer for these programmes will be a continuous improvement process, probably somewhat like the programme for the LCAC at Bell Halter in New Orleans.

Austal were awarded a contract for the first JHSV, the "Spearhead" for the US Army in November 2008, construction for this craft was started in December 2009 at their Mobile, Alabama plant. Delivery is due end 2011. The second vessel "Vigilant" for the US Navy was authorized in December 2009, and construction started in September 2010 for delivery in 2012. Authorisation has also been given for a further three vessels out of the total of the planned ten vessels to be constructed. The remaining five vessel construction contracts are due to be authorized over the period to 2013, achieving the objective of high-speed logistical support to the US Navy, Army and Marines. Key data for the Westpac and JHSV is given in Table 6.3.

The interesting aspect for these new military programmes will be whether there will be read across to ferry designs in the next decade. Both the LCS and the JHSV will be used quite intensively in rough conditions which should give feedback to improve motion damping systems. Machinery durability could also benefit from feedback from heavy usage. One clear difference to the LCAC programme is that there can be direct read across, whereas this has not really been possible from the LCAC.

Future Prospects

The military examples above are characterized by a significant competitive prototype trials programme followed by concept selection (or portfolio of concepts), and then detailed design for series production, followed by installation in the fleet including the base facilities, maintenance planning/refit programme, etc.

In contrast to this the application of fast craft as ferries has often been by an operator taking a risk, based on potential to build a new customer base, and then either having success, or quickly closing down. The difference between success and failure has been the operator's knowledge of terminal requirements, setting up robust trip timetables rather than being optimistic, and attention to detail. Many fast craft ferries in their initial versions were somewhat "experimental", pushing technical boundaries, and so unplanned events would happen, and these could either be

Fig. 8.24 Solent express at Ryde

accommodated or become a "problem" that turned passengers away. Care with planning and careful application of craft capabilities has been a hallmark of Hovertravel for example, and this is evident in their continued success after 36 years of service between Southsea and Ryde across the Solent, see Fig. 8.24.

Markets also change as society develops. In the 1950s, when the first hydrofoils became so popular in Italy, passengers were the market. Car ownership was not so ubiquitous then. Since the 1980s, particularly in Europe many routes saw a growth in demand for vehicle transportation, initially with traditional monohull ferries, and once the catamaran showed that it could accommodate large numbers of cars economically, this market took off. This has since been repeated in other parts of the world.

In the Mediterranean, monohulls and catamarans have developed competition with hydrofoils on many routes around Italy, Corsica, France, Greece, the Adriatic, and Turkey. The longer routes a number of these operators run were the challenge taken by the HYDAER programme to develop a passenger/car ferry that could operate above 100 knots and so make a step change in such services. Transformation of the concept into a ferry was challenged by fuel prices once again, as well as the conundrum of trying to stabilize sea skimming craft with both hydrodynamic and aerodynamic components; so performance at this level is still to be conquered.

Currently (2011), another world economic cycle is going through a dip, and so economics is being tested again as passenger numbers reduce for a while; operators look at extending the life of their vessels, and costs increase rather than decrease. Fortunately, in current times, the overall world economy means that travel has become a much more important part of people's lives and so the main result of the current cycle should be higher efficiency in all parts of the global transportation network. For the HPMV, this will lead the drive to higher efficiency power plant and

Fig. 8.25 Oman catamaran "Shinas" and Oman WPC ferries

higher safety levels. The widening market should help support concept development through business volume providing R&D funds.

While the economic dip has been significant for the Western world, it has been less so for the Middle East and China. In the Middle East a number of fast ferry services have been started up along the Western Gulf Coast, and also the Indian Ocean coast of Oman. These are all based on using catamarans and WPC. Austal has had success with their designs for Saudi Arabia (two AutoExpress 69 Jazan and Farasan operating from Jazan to Farasan on the West coast) and also Oman (two AutoExpress 65 Shinas and Hormuz operating along the Musandam peninsular coast from Muscat), see Fig. 8.25. The "Shinas" has recorded speeds of 56 knots on trials. The Omani government have also procured two WPC for coastal patrol purposes from Rodriquez to a design from AMD in Australia in 2008, see Fig. 6.10. These are aimed at patrol in the Indian Ocean which can be quite rough at times. This has been followed with an order for three WPC ferries of similar design from Rodriquez, the Sawqrah, Al Hallmayat, and the Masirah. The last two craft will operate to islands on the Omani south west coast in the Indian Ocean, while the Sawqrah operates services from Muscat to Khasab from its introduction in 2011. The outfit is a little special for these Omani craft from other routes as the craft have access for helicopters, a small hospital suite and overnight berths for the crew, as well as high quality passenger accommodation for 116, and vehicle capacity up to 22 cars.

Clearly the combination of straight ferry duties and the ability to perform as a rescue vessel are important to Omani Authorities. Where the passenger demand is not high and there is such a mission available this combination can be the best economic solution. In this case, the national authorities are developing the transport infrastructure via the fast ferries.

Fig. 8.26 Chinese WPC coastal patrol craft

In Oman, the Australian designers AMD have been able to team with Rodriquez, to build their WPC catamarans for ferries and utility missions. In China there are also WPC utility craft in service as shown in Fig. 8.26.

This is fine for the large commercial and military HPMV, but what about all those other applications? Most of them are rather small niches in the larger business context, and tend to be robust so long as the product is attractive. Will the market expand for such craft?

The *ACV* as a concept has stabilized as a specialist product for mainly utility and paramilitary missions. The cost is high but to achieve the same with alternative options would be even higher cost, or simply not possible. An example would be coastal patrol in very shallow waters, or icebound waters. The alternative of a helicopter is much more expensive and would not have the same work delivery as well as demanding highly trained personnel for operation and having a much more limited operational envelope constrained by wind and visibility. ACV operation does require training, but is not as costly as aerospace training.

The economy of the patrol ACV has lead to coastguard and naval forces in the Middle East, India, Sweden, Finland, and a number of other countries building small fleets of these craft to use in coastal areas that are difficult to access for conventional patrol boats.

The basic configuration for ACV's in their useful operating envelope will continue to be optimized, with most of the changes being internal. In this respect it is useful to consider the cars we drive and compare them with those of 50 years ago—apart from style they are still approximately the same size with the same basic attributes. Nevertheless, today's car has a wealth of electronics internally, suspension

Fig. 8.27 Super-yacht catamaran

that is highly effective and tuned, an engine that is very much more efficient, and a structure that has become highly effective to protect the occupants. It is this same kind of journey that the ACV and other HPMV's are going through.

The *SES* is till finding success both in commercial and military operations, as many of the SES produced by Brodrene Aa are still in service and the Mine Countermeasures vessels and Fast Attack Craft are in active service with the Norwegian Navy. It is not certain what the next generation of technical advance may bring for this concept, but it still remains the concept able to economically deliver the next level of speed above the semi planing catamaran.

At present, the *Super-yacht* has gone through a development in a similar way. Speed has increased from that of the "Victoria and Albert" of the nineteenth century but the focus is still on the facilities. It is a wonder at present why we do not already have trimaran super-yachts, as the potential for facilities on board will be even better than a monohull. There is still the issue over berth availability for such yachts, but this should reduce over time. The main issue is just that any super-yacht is a special project, the ultimate niche product.

One example of the trend starting to emerge for clients to go for the catamaran form is the 50 m catamaran super-yacht by Sabdes Design, of Melbourne, see Fig. 8.27. The company have taken their monohull 50 m design and split the hull to form an asymmetrical catamaran giving significantly larger accommodation for the smaller waterline length, rather than the client seeking a larger monohull super-yacht. The vessel is designed for up to ten passengers and speed in the 20 knot class. It will be built at a Taiwanese shipyard. The client has also had a communication suite designed in so that his yacht can become a "satellite headquarters". It will take a while for the impressive lines of many monohull super-yachts to be developed on the catamaran platform, but it will surely come as the opportunity for even grander entertainment facilities is realized by designers and clients alike. The Sabdes design is already close!

For such vessels the docking facilities do limit their operations, though if the catamaran design means a client mission can be delivered with a shorter vessel perhaps this may not be such a challenge. Commercial operators and the military are used to setting up dedicated docking facilities for the vessels as part of their investment, so this is less of a constraint. For the individual or charter craft owner it simply needs an analysis of the berthing locations with potential and perhaps small modifications to the vessel design to make the concepts attractive. On the other hand the catamaran does not have the best motion characteristics in cross seas, and so further analysis of guest occupation and cruising routes will be necessary to identify the target ownership market, together with further optimisation of the stabilization systems now fitted to many of the larger catamaran ferries, including T foils under the bows, midships stabilizers and stern stabilizers. The final consideration, for builders at least, is whether they are interested in building a bespoke craft to very high specification for a demanding client, with little prospect of repeat orders. Perhaps the construction community for catamaran HPMV is just now reaching the point where such bespoke projects can be considered worthwhile. In contrast there are a significant number of shipyards who can build a monohull superyacht, giving the buyer a competitive market to procure from.

Considering competitive *Sport* for HPMV we have already reached physical limits, probably more than once in the last century. Whether monohull, catamaran or multihull these craft are now largely limited by their ability to run with steady enough aerodynamics so that when bounced on a "rogue" wave they don't immediately submarine or flip nose over tail. In some ways the WIG is the ultimate result of that search, but their positive flight height means that they are not racing boats. Perhaps the development for racing craft will also be refinements internally, with more efficient engines, and continuously refined control systems, focused on the aerodynamic components of the boat while not crossing the line to a WIG form. One thing we do not have to worry about for such craft is funding, though the mission capability is important, focused as noted on the extreme speed and environment.

One area that the competition boats have conquered is propulsion with the use of "surface drive" propeller see Fig 4.13. While operating at lower efficiency than subcavitating propellers or water jets, these super ventilated propellers can provide reliable propulsion in the 100–150 knots speed range. Presently water jet systems are really only suitable in the speed range up to 50 or 60 knots due to the problems of turbulence and lower efficiency at very high speeds for units that are also able to function well in the low speed range. R&D at the major water jet manufacturers is pushing the boundary forward, but the efficiency level required for ferries to be economic is will be a challenge for speeds significantly higher than 70 knots.

The *Hydrofoil* has tested its speed envelope mainly through military prototypes. The core successes of this type have been the medium speed craft from Rodriquez/Supramar that proved very economic, and more recently the Jetfoil that has achieved similar success in Hong Kong and Japan with enduring results. The recent R&D through the EU funding for HYDAER has not yet created a breakthrough to higher

Fig. 8.28 Rodriquez FSH 38 hydrofoil

speeds, though Rodriquez new submerged hydrofoil vessel, the FSH-38 (Fig. 8.28) may show the way to a new economy for these same passenger routes for the future. The foil developments by Supramar [3-11], when combined with Rodriquez new overall configuration really could deliver the next step forward in efficiency for the fast passenger ferry.

The other multihull and hybrid vessels provide advantages in very specific circumstances and so their market will by nature be niches. We will continue to see new hybrid concepts developed, and where they can prove reliable, they will be adopted.

Will passenger transport, utility, or military markets cause yet more HPMV concepts to be invented? The continuing proposal of hybrids probably suggests that we have employed the range and combination of individual support forces that are practical already. The Hydaer programme experimented with a combination that had not been studied before and produced some very useful results. The other main combinations have all been experimented with already.

The market itself may also further widen—one example is providing a logistic service to offshore wind farms as we have referred to in Chap. 6. The design of installation vessels for these pylon mounted wind generators has already become a significant segment for the smaller shipyard community. In the case where the windmills are installed in shallow water outside major estuaries the use of ACV or shallow draft catamarans for maintenance logistics clearly offers an opportunity to optimize personnel deployment. We have given some sample data for such craft in Chap 6. Currently, the specification for such craft is around 25–30 m vessels, with a dual requirement for high-speed deployment, and stable motions while moored at site, which is driving development of SWATH and semi swath hull forms. Coastal offshore wind farms are being built at a very high pace through the period from 2008, and this will continue on a worldwide basis, so this marine market is also set to expand rapidly.

In the northern arctic, realistically there will be a significant change in the Arctic Ocean environment over the first half of the twenty-first century. The hydrocarbon resources in this same area are also too valuable to simply be left in place indefinately, whether considered as a carefully managed energy resource or as a source of funds for national development through exploitation. So environmentally acceptable ways will be found. The ACV and specialised HPMV have potential to be part of the solution as logistic support to such developments in the coastal areas at least will be a challenge that could be met by ACV's and other HPMV depending on the winter ice cover behaviour.

At the other end of the scale, there are two potentially exciting developments. First, there is the further development of the ocean going catamaran and trimaran as the current procurement by the US Navy matures and the potential for low cost aviation remote deployment becomes clear as an alternative to the current aircraft carrier fleet—reducing the cost and risks of deployment for the armouries of remotely operated planes in use by the USA for example. Second, there is the trans-ocean delivery of high value low weight cargo.

The *WIG* is still the potential solution looking for a problem in this respect, as evidenced by Boeing's studies some years ago that have not yet been translated into hardware. Maybe it will not in the end be Japan to US west coast that will create the demand, but rather China or Korea to Singapore or India to Japan with industrial Business to Business part finished product deliveries for final assembly on a just in time basis.

The question here is whether there is a space between current cargo delivery by container ships delivering very high volumes in the 15–30 knot speed range resulting in delivery times between a week inside Asia or 3 weeks to Europe or America, and air freight that can be delivered to most places within a week accounting for transit to and from the airport hubs used.

In this respect, the port facilities and collection or distribution service may have a significant influence on developments. If one considers the security issues at a typical large hub airport, and the fact that they are essentially dedicated to passenger traffic, with cargo as a niche, it would seem natural that an extension of marine port facilities could offer more efficient access and handling for cargo, as long as the containerization and vehicle berthing/loading can be refined more along the lines of air freight containers for compact high value freight.

In some ways, setting up a system for high value freight is more difficult than passenger ferry operations, as the shipper really needs to be able to offer a range of destinations from a hub and to be able to minimize the influence of the transit for loading, and final delivery as otherwise the benefit of the higher speed delivery might be lost. The HPMV can only be successful, therefore, if it is part of a complete system that delivers a satisfactory result.

At present the range available for an HPMV places transoceanic missions outside the current practical envelope, typical craft ranges are in the region between 200 and 1,000 nautical miles before refuelling at present. In contrast the "Sea fighter" has a maximum range of 4,400 nautical miles, enough to cross the Atlantic though not at its Gas turbine powered maximum speed of 50 knots—it would need to use its cruising diesels and complete much of the journey at a more relaxed pace of 20 knots. The diesel powered catamarans of Austal and Incat make their trans-global delivery journeys successfully in carefully managed stages with additional fuel storage on board. This kind of experience has been the background to the Transformable Craft approach for the US Navy.

This turns us back to the efficiency of power plants of course—the new development programmes starting up in Europe to deliver new "lean burn" diesel engines for trucks should be able to be fed into the next generation of large catamarans and trimarans and lengthen the range somewhat. The twenty-first century is going to be all about the transition from fossil fuel and our current power plant to alternative fuels and power plants. Incat is already moving forward with their design for an LNG fuelled craft, referred in Chap. 6, to reduce emissions.

Gas turbines are less fuel efficient than diesels, but when installed in an HPMV capable of 150 knots or so, their power density is sufficient to enable large craft. The vessel configuration and hull geometry needs further development from today's concepts, but with structure technologies following parallel paths to that for medium size craft—carbon fibre reinforced resins—it should be possible in the medium term to produce craft than can deliver economy—then we will be back to solving the terminal and total delivery issues!

Maybe the authors are still too optimistic in this. Time will tell!

Appendix 1
UK Military Hovercraft Trials Units

Background

When Christopher Cockerell was seeking support for the development of the hovercraft principle, it was the intervention of the late Earl Mountbatten that was of significant help. With his background in Amphibious Warfare, he immediately saw the potential for this new concept, so much so that it was initially classified as Top Secret.

Eventually, the Saunders Roe, SR.N1, was built and launched on 11 June 1959 to further evaluate the potential of the concept. This new vehicle attracted great publicity and interest in the UK. Military interest led to the formation of an Interservice Hovercraft Working Party in 1961 and the formation of the Interservice Hovercraft (Trials) Unit (IHTU) at HMS Ariel (later HMS Daedalus), at Lee-on-the-Solent just west of Portsmouth harbour.

IHTU personnel were drawn from the Royal Navy, Royal Marines, Royal Air Force and the Army. The different Service backgrounds and training assisted both in routine maintenance and fault finding on these special craft. Additionally, after the traditional 3-year tour, personnel were drafted to active units and able to spread the message about the usefulness of amphibious hovercraft.

Figure A1.1 shows the Unit in the NHTU days with SR.N6, BH 7 and VT 2 craft present.

The Early Days

In order to evaluate hovercraft military potential, to start with craft were hired from their manufacturers and operated from Lee-on-the-Solent. These evaluations served the double purpose of enabling Service personnel to gain experience of hovercraft operations and assisting manufacturers in the development of their craft.

L. Yun and A. Bliault, *High Performance Marine Vessels*,
DOI 10.1007/978-1-4614-0869-7, © Springer Science+Business Media, LLC 2012

Fig. A1.1 HMS Daedalus—Base for Naval Hovercraft Trials Unit

The first major evaluation was conducted with the SR.N1, at the time fitted with 18 in. skirts. Trials comprised many circuits of an overland hover way, which included steps and waves constructed out of shingle. In a second series of trials, this time with 4 ft 6 in. depth skirts fitted, the craft was operated over saltings with gullies nearly 3 ft deep and between 4 and 15 ft wide. SR.N1 was an experimental design, but SR.N2 was a practical craft and military trials included a passage from Cowes to Portland in seas up to 6 ft from crest to trough and a circumnavigation of the Isle of Wight in calm conditions, at an average speed of 58 knots. IHTU were also involved in demonstrations of SR.N2 in Montreal, operating on the St. Lawrence River and the Lachine Rapids.

Vickers Armstrong also produced early hovercraft and the VA 1 was based at Lee-on-the-Solent, carrying out obstacle clearing trials over straw bales on the airfield. The larger VA 2 was taken to RAF El Adam, in North Africa to conduct trials in desert conditions. The first ever comprehensive overland trials with a hovercraft saw VA 2 complete journeys totalling nearly 200 miles at speeds up to 38 knots. Skirt wear over the sand was severe, but neither dust, nor the terrain were problems.

It became obvious from the evaluations of hovercraft hired from their manufacturers that in order to fully evaluate hovercraft military potential, craft owned by the Unit would be needed. To this end, delivery was taken of the SR.N3 and three SR.N5s from the British Hovercraft Corporation in July and September 1964, respectively.

Fig. A1.2 SR.N5 in Sumatra Jungle

The Interservice Hovercraft Unit (Far East)

Two of the SR.N5s—XT 492 and XT 493—were military versions, modified for operations in the tropics, to be used by IHU(FE). After a period of training in the UK, the Unit moved to Singapore, at the end of 1964, for a 2-month work-up period (Fig. A1.2).

The Unit deployed to Tawau, Sabah, to assess the value of hovercraft as a general military load carrier in the type of terrain encountered in the area (Fig. A1.3). Logistic support was normally carried out by road vehicles, civilian river craft, helicopters or assault boats. Road communications in the area were extremely limited, river craft were hampered by the profusion of floating debris and by shallow water at low tide, and the use of helicopters was strictly limited, in order to conserve hours for more important tasks. However, at Tawau, the craft covered nearly 6,400 nautical miles and carried 1,600 passengers and 110 tonnes of freight.

The craft were moved to Thailand, where their amphibious capabilities were demonstrated, both in calm conditions and winds up to 40 knots, seas up to 7 and 5 ft surf. The craft were driven up and down river rapids flowing at speeds in excess of 12 knots, with rocky outcrops and through gaps not much wider than the craft. Tight turns in the rivers were often negotiated by using the inflated skirt as a fender against the river banks.

Fig. A1.3 SR.N5 at sea

Fig. A1.4 SR.N6 in Falklands

In Sarawak, a journey of over 300 miles along the Rayong River was completed. The first 110 miles were over open river, but from then onwards, commencing with the Pelagus Rapids, the route became increasingly difficult. The journey normally takes 8–12 days by long boat—the hovercraft completed the journey in less than 8 h.

It should be remarked here that the Far East Unit was successfully maintaining gas turbine-driven craft with aerospace structures and early generation skirts to operate in high humidity equatorial environment, making journeys into extreme remote locations.

Such performance did not go unnoticed in the Navy or Royal Marines, and craft were sent on trials or demonstration deployments to many challenging locations worldwide in the late 1960s and early 1970s, including detachment of a Navy SR.N6 to the Falkland Islands for a summer season tour in late 1967 (Fig. A1.4)

Trials and Evaluations

The SR.N3—the World's first military hovercraft—undertook a series of role evaluations, including operating as an RAF marine craft, anti-submarine warfare (ASW) trials for the Royal Navy and operations in the logistic role for the Army. The craft also featured in the first shop window for hovercraft—Hover show 66. The SR.N3 and later the BH 7 featured in trips abroad, but as with early commercial hovercraft ferries, the smaller SR.N5 and SR.N6s proved to be the "work horses" of the Unit.

SR.N5 XT 657 was fitted with a comprehensive instrumentation package to record trials data during performance trials, conducted by Royal Aircraft Establishment, Bedford and Admiralty Experiment Works, Haslar personnel. Bedford trials included static tests to measure propeller thrust, control forces and pitch and roll stability, lift system efficiency, hover height and daylight clearance. Sea trials included performance measurements to enable the variation of drag components with forward speed to be obtained, control effect investigations and seakeeping trials at three different weights and in a number of environmental conditions.

On return from the Far East, XT 493 was refurbished, and after employment on training and general logistic tasks, was prepared for surf trials. These were conducted at Saunton Sands on the North Devon coast, also involving an SR.N6 Mk 2. It was found that the best method to approach the beach was to position the craft between surf crests and adjust speed, until the craft and surf approached the shore at the same speed. Leaving the beach proved to be best on a track at 45° to the surf.

A military version of the SR.N6 conducted naval trials at Portland, investigating use as an anti-Fast Patrol Boat training craft, a helicopter safety boat and various exercises with ships working up with Flag Officer Sea Training. A major trial involved investigation into the best geometry for an overland "hover way" conducted at Longmoor, Hampshire, culminating in the construction of a 0.83 mile long hover way, with curves, ascents and descents, completed by the craft at the target mean block speed of 25 knots.

British Hovercraft Corporation, BH 7

The larger BH 7 completed a very comprehensive series of acceptance trials, operated by a joint team from IHTU and the British Hovercraft Corporation. Once accepted into service, crews were worked up to operate and maintain the craft.

Cold weather trials were conducted during March and April 1972, in Sweden and the Gulf of Bothnia. The journey of some 1,600 nautical miles to Sweden and on to Ranea, the northernmost navigable point of the Gulf of Bothnia, was one of the longest journeys undertaken by hovercraft operating under its own power.

Operating so far from base, in relatively harsh conditions and over surfaces not previously encountered, presented the craft and crew with a major challenge. This challenge was met and the trials successfully completed. The results attracted significant interest, particularly the demonstrating of ice breaking by hovercraft.

Evaluation of long-term habitability and reliability was tested when the BH 7 was used on Channel Patrol, operating from Pegwell Bay, Kent. The craft was directed by HM Coastguards at St. Margaret's Bay, Dover, usually while underway in the Channel, the craft being dispatched to identify a "rogue" spotted on the Dover radar screen. Interceptions were made by day and night, the object being to locate and photograph the "rogues". Interceptions were carried out in weather conditions up to 20 knot winds and 4 ft significant seas.

In the latter half of 1973, the BH 7 was deployed to the East Coast of the USA, the craft being fitted with a clam shell bow door and ramp to enable vehicles and large loads to be carried, prior to departure. The majority of the work in the USA consisted of presentations to both military and civilian authorities. Trials were conducted for the Naval Ship Research & Development Centre and exercises conducted with the US Marine Corps and the US Coastguards. The craft attracted several thousand visitors when in New York.

Mine Counter Measures

In mid-1971, the emphasis of the Unit trials shifted from general evaluations, performance and Army support to the execution of practical work to investigate the use of hovercraft in the Mine Countermeasures (MCM) role. This initially involved SR.N6s and later the BH 7 towing actual MCM and simulated MCM equipment, in a range of sea states. Equipment was developed both to enable the hovercraft to tow and control the streamed gear, and also to enable the craft crew to deploy and recover the gear. The attractiveness of hovercraft to the MCM role centred on their relatively low acoustic, magnetic and pressure signatures, relative immunity to underwater explosions and high transit speed potential.

SR.N6 MCM Tasks

Another associated area investigated was the recovery of "disabled" vessels. An SR.N6 was used to tow a 60 tonnes RPL and a 350 tonnes Coastal Mine Sweeper. These vessels were towed along straight courses on various headings relative to the wind and during turns. The ability of SR.N6 to tow SR.N3 and BH 7 was also demonstrated.

One of the in service precursor sweeps was the Pipe Noisemaker (PNM), which consists of three vertical pipes contained in a frame, suspended underneath a float. Tow lines are attached to the tube frame and the float, and as the equipment is towed through the water, the inner tube oscillates between the two outer tubes, resulting in a radiated noise pattern.

A sweep developed specifically to deal with mines that might be set for hovercraft—antennae mines—was the snag line sweep, operated by two SR.N6s. The equipment towed by both hovercraft consisted of a float, underneath which was suspended a triangular plate and lead ball. The towline was attached to the triangular plate and the sweep wire ran between the two plates. Special equipment was developed at the Unit to enable the two hovercraft to maintain the required distance apart and the required position relative to each other. This sweep was used in the National MCM Exercise Scotch Broth, conducted in the Clyde Estuary in early 1973.

BH 7 Activity

The BH 7 was also heavily involved with the MCM evaluation, towing two US helicopter sweeps—the Mk 103 Mechanical sweep and the Mk 104 Acoustic sweep. The US Mk 103 sweep consists of two sweep wires, to which are connected the necessary otters, kites, depressors, floats and cutters; the resultant configuration streaming in the shape of a Vee behind the lead float. The 103 could not be deployed or recovered from the BH 7, so initially a transfer technique was involved, using a conventional minesweeper. At a later date, a sweep deck was added to the craft and deployment and recovery made possible without external assistance. The US Mk 104 is an acoustic sweep which can be handled using a small derrick.

National MCM Exercises

A natural development of the investigation into the use of hovercraft in the MCM role was participation in national MCM exercises. These are held in various locations around the UK coast and are designed to test the operational capabilities of MCM forces.

The hovercraft participation provided an opportunity for IHTU to demonstrate techniques developed during trials, under realistic operational conditions. Also, those associated with conventional MCM vessels were able to witness the hovercraft's capabilities and administrative officers were able to gain experience in the best utilisation of hovercraft, in conjunction with conventional vessels.

The chosen locations for these exercises were those most likely to be mined by enemy forces, in the event of hostilities. They are either harbours which are involved with the receipt and dispatch of goods, or more strategically important areas. An example of the former would be the Humber Estuary and adjacent areas, which were on the BAOR supply routes. The Clyde Estuary, which is the route by which submarines leave or return to their base at Faslane, is an example of the latter.

The hovercraft involved in the MCM exercises has operated from temporary bases, chosen during reconnoitre visits to the area. During these visits, discussions will have been held with the appropriate civil and service authorities and the required facilities organised. In fact, these facilities were minimal and are mainly domestic, as the Unit took adequate spares and equipment to enable most problems to be overcome.

The pattern for such exercises is very similar in that having made arrangements for the detached operating base, shortly before the exercise is due to commence, an advance party is deployed. This party sets up the communications and control centre, maintenance areas and mobile hovercraft operating base (MOHOB). The latter consists of custom-modified caravans which provide temporary accommodation on site and a maintenance control centre.

The craft taking part in the exercise will then arrive, either under their own power, or onboard ship transport. After an area familiarisation and work-up phase, the exercise proper commences. Practice mines will have been laid along the designated routes to be swept and it is the task of the conventional minesweepers and hunters and hovercraft to clear these routes. The exercise is discussed at a debrief, in order that any lessons learnt can be promulgated while impressions are still fresh. The final report will usually carry an analysis of the effectiveness of the exercise.

Naval Hovercraft Trials Unit

The Interservice Hovercraft Unit, as it was then known, was disbanded on 31 December 1974, consequent on the withdrawal of Army and Royal Air Force support, although the disbandment ceremony took place on 19 December. The ceremony was simple, consisting of a parade of service personnel and inspection by Brigadier R M Perkins, Chairman of the Defence Hovercraft Committee.

The original intention had been to hold the commissioning of the Naval Hovercraft Trials Unit (NHTU) close to the disbandment, but it actually took place on 17 January 1975. Guests at the ceremony included Frank Judd (Under Secretary of State for the Royal Navy), Vice Admiral P M Austin RN (Flag Officer Naval Air Command), the Chaplin of the Fleet, Capt A P Comrie (Captain of HMS Daedalus) and representatives from Government establishments and hovercraft firms, and the man "who started it all"—Sir Christopher Cockerell.

The trials executed by NHTU were mainly geared to the MCM hovercraft investigations and involved continuing trials on BH 7 and on SR.N4s and the Vosper Thornycroft, VT 2, hired from their operators and manufacturer, respectively. The VT 2 was later purchased by the MOD and converted for the logistic support role.

BH 7/SR.N4 Trials

Interest in acquiring hovercraft motion data increased in early 1975, in both the UK and the USA. This led to a joint operation at Pegwell Bay—Hoverlloyd's base, involving BH 7 and an SR.N4, data being recorded during commercial crossings of the Channel. Both craft were fitted with a US Navy data acquisition system, designed to be easily installed; data being recorded on a four track tape recorder. The object of the exercise was to collect accurate motion data experienced by the two craft, while operating in identical sea conditions.

As part of the MCM investigations, a contract was awarded to the British Hovercraft Corporation to conduct a series of model tests on various large hovercraft, including BH 7. These tests established the variation of drag with forward speed and craft yawing moments and side forces at small yaw angles, in calm conditions and 2½ ft regular waves. The full-scale trials involved towing various drag loads from a central and offset towing point; the results showing correlation was reasonable.

Similar trials were conducted with an SR.N4, operating from Pegwell Bay. In addition and in common with IHTU/NHTU craft, an SR.N4 was operated on the noise range at Portland.

Vosper Thornycroft, VT 2

The VT 2 was the largest hovercraft in the NHTU fleet and initially carried out trials to evaluate manoeuvring, control and seakeeping, with particular attention being paid to the characteristics of the large ducted propulsion fans and the non-compartmented skirt. These evaluations were conducted with the craft in free-running condition and when towing loads consisting of drogues or 12 in. circumference manila rope. As with the SR.N4, the acoustic signature was measured on the range just outside Portland Harbour, in early 1976.

For the logistic support role, the craft was fitted with a central roof hatch and bow door and ramp. Stores, personnel and equipment would be loaded into the central bay, via the bow ramp and door. The craft would then transit to the MCM fleet at high speed, come alongside and the stores, etc., loaded onto the conventional vessels by using the ship's crane.

VT 2 participated in a number of national MCM exercises and circumnavigated the UK under its own power—possibly the only hovercraft to have done this.

Other Activities

A task made easier by using a naval SR.N6 was the clearing of unexploded bombs and rockets on the wide mudflats off the Essex coast. The unexploded devices were left after bombing ranges ceased to be used, while others are relics from wartime accidents, or aircraft ditching. Although most of the bombs are inert practice devices, some are live and consequently highly dangerous. Members of the Portsmouth and Medway Bomb and Mine Disposal team worked from the hovercraft in January 1978 (Fig. A1.5). The craft were kept running and live munitions detonated with simple delayed action fuses rather than by wired remote. Large areas were cleared in days that would have taken months otherwise.

Fig. A1.5 SR.N6 bomb clearance

In the late 1970s, two naval hovercraft were deployed to Hong Kong, to help deal with the illegal immigrants coming from mainland China. These immigrants often used high-speed boats that the traditional police marine vessels had difficulty in catching. However, the hovercraft, with the ability to take a straight line route through shallows and mudflats, was more than a match. The junks used to transport the immigrants often came at night, with no navigation lights, but with the SR.N6 on radar watch, interception was possible, arrests being made by a Hong Kong policeman carried onboard. Transits were made at 25 knots reasonably safely. At 50 knots, extreme care had to be exercised, in the light of the general congestion in the Harbour.

NHTU Disbanded

The Ministry of Defence announced in late March 1982 that the NHTU was to close and its assets sold. Reasons given were partly to save costs, and partly that the Unit had carried out as much assessment as it could. The 100 or so personnel who manned the Unit were dispersed to other postings during the summer of 1982, while the hovercraft were disposed of in due course. The BH 7 was operated on a joint basis with the British Hovercraft Corporation for a while, who were embarking on an export drive for sales with the craft mainly to the middle-east including Iran and Saudi Arabia.

Fig. A1.6 VT 2, BH 7, and SR.N6 in formation in front of NHTU

The MOD stated at the disbandment that a hovercraft design was still among those under consideration for a new class of Mine hunter, the role which had been a major part of NHTU's evaluations in the last few years. All the NHTU craft had been refitted for some aspect of Mine hunting, or logistic support to MCM operations. Figure A1.6 above shows three NHTU craft, VT2, BH.7 and SR.N6 in a fly-past of the HMS Daedalus entrance ramp. Since that time, the UK Navy has focused more on blue water missions, while the Royal Marines procured a squadron of amphibious hovercraft from Griffon that are used for amphibious assault missions and have proved very effective during deployments in Africa for example.

Glossary

ACC	Planing monohull craft (PMH) + air cushion layer = air cushion cavity (ACC)
ACV	Air cushion vehicle
AR	Aspect ratio
α	Hull dynamic trim in pitch
β	Deadrise angle
B	Wing span
B	Hull waterline breadth
C, c	Wing Chord
CATB	CAT with bulbous bow
DACC	Dynamic Air Cushion Craft
DACWIG	Dynamic Air Cushion (PAR) WIG
delta h	water surface depression
FACAT	Catamaran (CAT) + hydrofoil (HYC) = foil-assisted catamaran (FACAT)
Frl	Froude Number based on waterline length $V/(gL)^{0.5}$
Fr_L	Froude number
GEZ	Ground effect zone
GRP	Glass reinforced plastic
H, h	Flying height
H	hours
h	Flying height of wing above still water
H/B	Relative flight height; flight height *H* over wing span *B*
H/C	Relative flight height; flight height *H*, over wing chord *C*
HPMV	High-performance marine vehicle, encompassing planing monohull boats, catamarans, trimarans and other multihulls, surface effect ships, hydrofoils, and wing in ground effect craft
HYC	Hydrofoil catamaran

HYF	Hydrofoil
HYSWAC	Small water plane area twin hull craft (SWATH)+hydrofoil (HYC)=hydrofoil small water plane area twin hull craft (HYSWAC)
ICAO	International Civil Aviation Organisation
IMO	International Maritime Organisation
kg	Kilogrammes, weight
kPa	Kilopascals, pressure
kph	Kilometres per hour
kW	Kilowatt, power
L	Hull water line length
L/B	Length/Breadth
LCAC	Landing craft, air cushion
M	Righting moment
NATO	North Atlantic Treaty Organisation
PAR	Power Assisted Ram wing
PCAT	Planing catamaran
PACSCAT	Air cushion vehicle (ACV)+catamaran (CAT)=partial air cushion supported catamaran (PACSCAT)
PAR	Power-assisted RAM
PS	Power, horsepower measured at output shaft of engine
S	distance to effective origin for air jet cone
SES	Surface effect ship
SEZ	Surface effect zone
SWATH	Small water plane area thin hull ship
T	Transformable (Craft)
tj	Air jet 'thickness'
TPC	Tunnel planing catamaran
Us	Maximum Velocity of air jet at its centre
WL	Water line
WW1	The First World War (1914 – 1918)
WW2	The second World War (1939 – 1945)
Xj	Distance to leading edge of wing

References

Chapter 1

[1-1] Willis S. The fighting temeraire. Quercus, London. http://www.MailOnline.co.uk

[1-2] Mudie C (1975) Power boats. Hamlyn Publishing, London. ISBN 0 600 37045 3

[1-3] Du Cane P (1974) High speed small craft, revised 4th edn. David and Charles, Devon. ISBN 0 7153 5926 6

[1-4] Yun L, Bliault A (2000) Air cushion craft. Arnold Publishers (now Hodder Headline, part of Elsevier), London. ISBN 0 34067650 7 (UK), ISBN 0 470 23621 3 (Wiley)

[1-5] OM Faltinsen (2005) Hydrodynamics of high-speed marine vehicles. Cambridge University Press, Cambridge, 451 p. ISBN 978-0-521-84568-7

Chapter 2

[2-1] Wheeler RL, Chaplin JB (2008) The SR.N1 hovercraft. Cross Publishing, Chale. ISBN 978 1 873295 17 5

[2-2] Newman JN, Pool FAP (1962) Wave resistance of a moving pressure distribution in a canal. Schiffsteknik 9:1–6

[2-3] Yun L, Sun Y (1988) Development of air cushion vehicles in China in the last three decades. In: First international HPMV conference, Chinese Society of Naval Architects and Marine Engineers, Shanghai, Nov 1988

[2-4] Zhou WL, Hua Y, Clao X (1984) Analysis of the stability of ACV. In: Second conference on HPMV, Chinese Society of Naval Architects and Marine Engineers, Shanghai, 1984

[2-5] UK Department of Industry (1980) Stability and control of hovercraft—notes for commanders. UK Department of Industry, London

[2-6] King HF (1966) Aeromarine origins: the beginnings of marine craft, winged hulls, air-cushion and air-lubricated craft, planing boats and hydrofoils. Putnam, London

[2-7] Wheeler RL (1993) From river to sea: the marine heritage of Sam Saunders. Cross Publishing, Newport. ISBN 1 873 295 05 07

[2-8] Russell BJ (1979) The interservice hovercraft (trials) unit. Hover, Gosport. ISBN 0 950 647 004

[2-9] McLeavy R (ed) (1973) Jane's surface skimmer systems: hovercraft and hydrofoils 1971–72. Jane's Year Books, London

[2-10] Dickins DF, Cox M, Thorleifson J (2008) Arctic patrol hovercraft: an initial feasibility study. In: IceTech 2008, Banff, paper 08-115-RO, 2008

Chapter 3

[3-1] Gibbs-Smith CH (1975) Early flying machines 1799–1909. Eyre Methuen, London. ISBN 978-0-4133-3020-8

[3-2] Rozhdestvensky KV (2006) Wing-in-ground effect vehicles. Progr Aero Sci 42. http://www.sciencedirect.com

[3-3] Heber CE (1981) The power augmented ram wing in ground effect concept as an airborne Amphibious quick reaction force. Paper AIAA-81-2077 presented at AIAA 6th marine systems conference, Seattle, Sept 1981

[3-4] Yun L, Bliault A, Doo J (2010) WIG craft and ekranoplan: ground effect technology. Springer Science and Media, New York, 441 p. ISBN 978-1-4419-0041-8

[3-5] Wong KP, Yun L (2002) An introduction to the novel WIG. In: Proceedings of international conference on HPMV, Shanghai, 24–27 Apr 2002

[3-6] Yun L, Guihua P, Xie Y, Sheng J, Wu C (2000) Research and design of the dynamic air cushion WIG (DACWIG) type 'Swan'. In: 3rd international conference for HPMV, Shanghai, Apr 2000

[3-7] IMO interim guidelines for wing in ground effect craft MSC/circ 1054 and MSC/circ 1162 2005

[3-8] IMO high speed craft code, document IMO-187-E (English). ISBN 0 86039 128 0

[3-9] Bucci A, Roccotelli S, Marchetti M (2000) Impact of semi-submerged obstacles on the composite fin of a new concept high velocity craft. In: International Society of Offshore and Polar Engineering, ISOPE 2000, Seattle, 28 May to 2 June 2000

[3-10] de Wolf WB (2002) Aerodynamic investigations on a wing in ground effect. Report NLR-TP-2002-506. National Aerospace Laboratory, Amsterdam

[3-11] La Roche U, Trevisani H (2001) Supramar cavitation suppression at hydrofoil thick sections reference paper: replacing cavitation and supercavitation bubles on profiles for high speed propellers and foil by two-phase boundary layers on profiles at high speed and low sigma. In: Second high-performance marine vehicles conference, 2001 Technical University of Hamburg-Harburg, Hamburg. http://www.supramar.ch/supramar_anti_cavitation_groove.htm, http://www.supramar.ch/index.htm.

[3-12] Gehling R (2010) High speed craft—uniform operating limitations. Int J Small Craft Technology, Trans Royal Inst Naval Architects Vol 152(pt B2). ISSN 1740-0964

Chapter 4

[4-1] Johnstone P (1980) The sea-craft of prehistory. Routledge and Kegan Paul, London. ISBN 0 7100 0500 8

[4-2] Morrison JS, Coates JF (1986) The Athenian trireme—the history and reconstruction of an ancient Greek warship. Cambridge University Press, Cambridge. ISBN 0-521-3100-4

[4-3] Various contributors (1975) The lore of ships. Nordbok, Gothenberg

[4-4] Howarth D (1974) Sovereign of the seas: the story of British sea power. Collins and Company, Glasgow

[4-5] Kemp P (1976) The Oxford companion to ships and the sea. Oxford University Press, Oxford. ISBN 0 586 08308 1

[4-6] Menzies G (2002) 1421: the year China discovered the world. Bantam Books, New York. ISBN 978-0-553-81522-1

[4-7] Greenhill B (1980) Schooners. B. T. Batsford, London. ISBN 0 7134 3380 9

[4-8] Dawson C (1972) A quest for speed at sea, the story of the Vosper shipyards founded by Herbert Edward Vosper. Hutchison, London. ISBN 0 09 109 720 7

[4-9] Hill-Norton L (1982) Sea power: a story of warships and navies from dreadnaughts to nuclear submarines. Faber and Faber, London. ISBN 0 571 11890 9

[4-10] RINA (2010) The Royal Institution of Naval Architects 1860–2010. RINA, London. ISBN 978-1-905040-74-2

[4-11] Comstock JP (ed) (1967) Principles of naval architecture. Society of Naval Architects and Marine Engineers, New York. Library of Congress Catalog 67-20738

[4-12] Clement EP (2006) A configuration for a stepped planing boat having minimum drag (Dynaplan boat). http://www.dynaplaneboat.com

[4-13] Lambert J, Ross AL (1991) Allied Forces of World War II, volume 1: fairmile designs and US submarine chasers. Conway Maritime Press, London, 256 p, many drawings and construction lists. ISBN 978-0851775195

[4-14] Lambert J, Ross AL (1993) Allied Coastal Forces of World War II, volume 2: Vosper MTBs and US ELCOs. Conway Maritime Press, London, 256 p, many drawings and construction lists. ISBN 978-0851776026

Chapter 5

[5-1] Barnaby KC (1964) 100 Years of specialised shipbuilding: the story of John I Thornycraft and his shipyards 1864 to 1964. Hutchinson, London

[5-2] Gunston B (1969) Hydrofoils and hovercraft. Aldus Books, London. ISBN 490 00135 1 and 490 00136 X

[5-3] Cagle MW (1970) Flying ships, hovercraft and hydrofoils. Dodd Meade and Company, New York

[5-4] MCleavy R (1976) Hovercraft and hydrofoils. Blandford Press, Poole Dorset. ISBN 0 7137 0767 4

[5-5] McLeavy R (ed) (1972) Jane's surface skimers: hovercraft and hydrofoils 1970–71, 4th edn. Jane's Yearbooks, London, p 182. ISBN 07-032203-1

[5-6] Foss B (1989) Hurtigbåten, Gammeldampen Arvtager. Nordvest Forlag, Ålesund. ISBN 82 9033046 4

[5-7] McLeavy R (ed) (1975) Jane's surface skimers: hovercraft and hydrofoils 1973–74, 7th edn. Jane's Yearbooks, London, p 245. ISBN 0-354-00116-7

[5-8] Rodriquez reports on fully submerged hydrofoil trials. Fast Ferry International, Apr 2010

[5-9] Bucci A, Palmieri D, Roccotelli S, Marchetti M (2000) Impact of semi-submerged obstacles on the composite fin of a new concept high velocity craft. In: International Society of Offshore and Polar Engineering Conference, Seattle, June 2000

[5-10] Pistani DDF (2002) An experimental approach for the prediction of full scale forces acting on the submerged control surfaces of a new WIG vehicle. In: International offshore and polar engineering conference, Kitakyushu, 2002. ISBN 1-880653-58-3

Chapter 6

[6-1] Neal C, Janus TK (2001) Puget sound ferries: from canoe to catamaran. American Historical Press, Sun Valley, 220 p. ISBN 978-1892724199

[6-2] Dag Bakka Jr (2005) Selskabe. Det Stavangerske Dampskibsselskab 150 år, DSD, Stavanger. ISBN 82 303 0571 4

[6-3] Dubrovski V, Lyakhovitsky AA (2001) Multi-hull ships. Backbone Publishing, Fair Lawn, 495 p. ISBN 978-0964431126

[6-4] Stena report on 18 months of HSS 1500 services. Fast Ferry International, Nov 1997

[6-5] Varndell M (2011) 'The British are coming'—a look at British record breaking and the people that make it happen. Transport Bookman, London. ISBN 9780851840765
[6-6] Austal Auto Express 102 trimaran trials programme completed. Fast Ferry International, May 2010
[6-7] Kennell C (1992) SWATH ships. Technical and Research Bulletin 7-5 prepared for Panel SD-5 (Advanced Surface Ships and Craft), SNAME 1992
[6-8] Dubrovski V (2007) Small waterplane area ships. Backbone Publishing, Fair Lawn, 255 p. ISBN 978-0974201931

Chapter 7

[7-1] Pavlov G (2002) The contribution of the Russian inventors to development of ships on artificial cavity (SAC). In: Proceedings of HPMV'CHINA2002, Shanghai, 24–27 Apr 2002
[7-2] Pavlov G, Yun L (2002) Development & performance of air cavity craft. In: Proceedings of HPMV'CHINA2002, Shanghai, 2002
[7-3] Pavlov G (2003) Development of air cavity craft in Russia. In: Proceedings of international annual conference for HPMV, Shanghai, 2003
[7-4] Sverchkov AV; Krylov Shipbuilding Research Institute (2010) Application of air cavities on high speed ships in Russia. In: Paper 11, international conference on ship drag reduction (SMOOTH Ships), Istanbul Technical University, Istanbul, 20–21 May 2010
[7-5] Full scale air supported vessel starts trials. Warship Technology, Oct 2010, published by The Royal Institution of Naval Architects
[7-6] Pavlov S, Yun L (2005) A brief introduction to a high speed passenger craft—hydrofoil assisted CAT, superfoil 40. In: Proceedings international annual conference for HPMV, Shanghai, 9–10 Apr 2005
[7-7] Brizzolare S (2003) Hydrodynamic analysis of interceptor with CFD method. In: /th Internationa Conference on Fast Sea Transportation (FAST 2003), Ischia Porto, Italy
[7-8] Hart CJ, Weems KM, Peltzer TJ (2007) Seakeeping analysis of the lifting body technology demonstrator sea flyer using advanced time-domain hydrodynamics. In: 9th international conference on fast sea transportation, FAST 2007, Shanghai, Sept 2007
[7-9] Gee N et al (2005) The X craft—a potential solution to littoral warfare requirements. In: Proceedings of HPMV international annual conference, Shanghai, 9–10 Apr 2005
[7-10] Harrison J (2006) Seafighter (FSF-1) ex-X-Craft, ex-LSC(X), myth and reality. In: Navsea 05D4 presentation, 28 June 2006
[7-11] PACSCAT technology demonstrator delivered. Warship Technology, Oct 2010, published by The Royal Institution of Naval Architects
[7-12] Clements RJ, Wilson PA, Lewthwaite JC, Molland AF, Ivanov P (2005) The potential for the use of a novel craft, PACSCAT (partial air cushion supported catamaran) in inland European waterways. In: International conference on fast sea transportation, FAST '2005, St. Petersburg
[7-13] Trials program of M80 test craft starts in California. Fast Ferry International, Mar 2006

Chapter 8

[8-1] Papanikolaou A (2005) Review of advanced marine vehicles concepts. In: National Technical University of Athens, Ship Design Laboratory, high speed marine vehicles conference, Athens, 2005

[8-2] Cooper K (2009) Sea base enabler innovative naval prototype transformable craft (T-Craft), SNAME SD-5 Panel Brief, Apr 2009

[8-3] Bodden HJ (June 2010) A survivability assessment of the transformable craft in an operational environment. Masters Thesis, Naval Postgraduate School, Monterey

[8-4] Lu SH, Yun L (2009) A high speed ferry of compound CAT. In: Proceedings of Shanghai HPMV conference, Shanghai, 2009

[8-5] Ge WZ et al (2009) A SWATH project for Taiwan Strait Ferry. In: Proceedings of Shanghai HPMV, Shanghai, 2009

[8-6] (2006) Taking back the littoral (US Navy Littoral Combat Ship Program Update and More). US Navy Force J III/2006

[8-7] Joint High Speed Vessel (JHSV) Program Overview. Presentation to SNAME SD-5 and International Hydrofoil Society, Sept 2006

[8-8] Burgess Marine completes rebuild of Incat 96m wavepiercer. Fast Ferry International, May 2009

[8-9] Information and further references on USS Freedom can be found at: http://en.wikipedia.org/wiki/USS_Freedom_(LCS-1)

[8-10] Frey C (2009) Joint High Speed Vessel Program, Program Overview. Presentation to SNAME SD-5 and International Hydrofoil Society, July 2009

Resources

We list below a small selection of publications that give regular information on HPMV and a selection of internet sites that may form a starting point for search. First the publications:

Fast Ferry International,
ISSN 0954 3988, published ten times yearly. Details of Fast Ferry HPMV and the services, vessel and operator directories. Go to the internet location http://www.fastferryinfo.com for details.

Ship and Boat International,
The Royal Institution of Naval Architects, London, UK, ISSN 0037 3834. Also Warship Technology published by the RINA, ISSN 0957 5537. Both Journals have articles on high-speed craft. Find the RINA at http://www.rina-org.uk.

Boat International,
Boat International Media Ltd, London, UK, ISSN 0264 9138. Monthly magazine focusing on superyachts and chartering. Go to http://www.BoatInternationalMedia.com for details.

Yachts France,
Lux Media Group, Cannes, France. Monthly Magazine focusing on fast motor yachts and super yachts. Go to http://www.luxmediagroup.com for details.

Sports Boat and RIB,
CSL Publishing Ltd, Cambridge CB2 3HX, England. Monthly Magazine focusing on sports powerboats including RIB. Go to http://www.sportsboat.co.uk for details.

The FAST Series of Marine Conferences
on Fast Sea Transportation, These cover design and technology research for all types of HPMV. The papers have become steadily more analytical as the conferences progress from their initiation in Trondheim in 1991. Go to FAST 2009 or FAST 2011 for more details.

Societies

The Societies that best form a start point are RINA, SNAME, AIAA and International Hydrofoil Society as in the internet list below, as these all publish research papers and technical journals. There are many other societies and special interest clubs that can provide useful information if a keyword search is used.

http://www.RINA.org.uk	Royal Institution of Naval Architects (London, UK)
http://www.SNAME.org	Society of Naval Architects and Marine Engineers (NY, USA)

http://www.AIAA.org	American Institute of Aeronautics and Astronautics
http://www.MARIN.nl	Site for Netherlands Test Basin
http://www.NTNU.no	Site for Trondheim University and Test Basin
http://www.foils.org	Site of the International Hydrofoil Society

A selection of sites on the internet that are useful for following up:

General

http://www.usn.mil	The site for US Navy
http://www.uscg.mil	The US Coastguard site
http://www.ONR.Navy.mil	Office of Naval Research, USA
http://www.dfdickins.com	Engineering and environmental research company
http://www.Rolla-propellers.ch	Rolla propellers (also analytical consultants)
http://www.Arneson-industries.com	Arneson surface drives
http://www.KHUP.com	Reference site for publicly available papers such as HSC code updates (search on keyword HSC-code)
http://www.royalnavy.mod.uk	Site for British Royal Navy
http://www.hmswarrior.org	HMS Warrior ship and museum in Portsmouth
http://www.theblueriband.com	History and ships' information of the Blue Riband Trophy
http://www.janes.com	Search on High-Speed Marine Transportation for further details

Hydrofoils

http://www.supramar.ch/index.htm	The site for Supramar ag
http://www.hydroptre.com	Site with info on large sailing trimaran hydrofoil
http://www.usshighpoint.com	Site for USS High Point alumni and refurbishment
http://www.rodriquez.it	The site for Rodriquez hydrofoils

ACV

http://www.acvdesigns.com	ACV Designs/Canair site
http://www.airlifthovercraft.com	Hovercraft builders in Australia for utility market
http://www.almaz.spb.ru	Shipbuilders of military fast craft including ACV and patrol boats based in St. Petersburg, Russia
http://www.bbvhovercraft.co.uk	BBV hovercraft site
http://www.Griffonhoverwork.co.uk	Griffon Hoverwork site
http://www.hovercraft.com	Universal Hovercraft site
http://www.neoterichovercraft,com	Neoteric Hovercraft site
http://www.revtechover.com.au	Revtec Hovercraft Site
http://www.Textronmarineandland.com	Site for Textron Marine working on T Craft and LCAC
http://www.um.no	Site for UMOE Mandal, working on ACV's and T Craft
http://www.viperhovercraft.com.au	Viper Hovercraft Site

Monohulls

http://www.triremetrust.org.uk	Site for the Trireme Trust (ref Olympias reconstruction and test)
http://www.superyachts.com	A site with data on super yachts and tenders
http://www.cigaretteracing.com	The site for fast racing boat builders Cigarette

http://www.windy.no	The site for windy boats
http://www.sunseeker.com	Sunseeker superyacht builders
http://www.seseu.com	Effect Ships International
http://www.aeromarineresearch.com	Information site for power boat design
http://www.boatdesign.net	Information network for boat design
http://www.vintagehydroplanes.com	Information exchange on vintage racing boats
http://www.lesliefield.com/	Hydroplane History including Harmsworth Trophy, speed records with various craft and the Campbells' Bluebirds and much more
http://www.Bluebirdsupportersclub.com	Site for history of the Campbells speed records boats
http://www.fincantieremarinegroup.com	Fincantieri Shipbuilders site
http://www.damen.nl	Shipyard constructing fast patrol vessels, supply vessels, and ferries
http://www.epsilonmarine.gr	Builders of FRP monohull ferries in Greece
http://www.fincantieri.it/	Fincantieri marine group builders of fast ferries, mega yachts, and cruise liners
http://www.danishyachts.com	Builders of high-speed superyachts
Catamarans and trimarans	
http://www.austal.com	Austal Catamarans
http://www.amd.com.au	Advanced Marine Designs—wavepiercer specialists
http://www.incat.com.au	International Catamarans site—wavepiercing catamarans
http://www.gdlcs.com	General Dynamics site for the LCS programme
http://www.QUINETIQ.com	Site for Pacscat (use search at home site)
http://www.bmtniggal.co.uk	Site for catamarans and X Craft
http://www.Bluebirdsupportclub.com	Site for details about Malcolm and Donald Campbell speed record craft

SWATH

http://www.swath.com	Site for SWATH International Ltd of Bethesda, MD
http://www.navships.com	Site for Navatek designers of SWATH and submerged buoyancy craft

WIG

http://www.wigetworks.com	Site for Singapore operator of FS-8 WIG
http://www.wingship.com	Site for Wingship Technology of Korea
http://www.cnsamt.com	Site for manufacturers of Aron-7 and other passenger WIG
http://www.attk.ru	Arctic Trade and Transport Company, builders of the Aquaglide 5 passenger Ekranoplan
http://www.seaeagleinternational.com	Builders of Sea Eagle WIG craft

Propulsion

http://www.wartsila.com	Wartsila water jets in range 4,500–26,000 kW
http://www.mjp.se	MJP water jets from Sweden
http://www.hamiltinjet.co.nz	Hamiltonjet water jets in range to approximately 4,000 kW
http://www.rolls-royce.com/marine/	Rolls Royce subsidiary KaMeWa water jets in power range from 100 kW to 40 MW (under/propulsors/waterjets)

Museums

The following museums are mentioned in the text, so readers may like the contact data, as follows:

Discovery Museum Tyneside:
Blandford Square, Newcastle upon Tyne, Tyne and Wear NE1 4JA, UK. http://www.twmuseums.org.uk

The Hovercraft Museum:
Building 40, Daedalus site, Argus Gate, Chalk Lane, Lee-on-Solent, Hampshire PO13 9JY., UK http://www.hovercraft-museum.org

US Army Transport Museum:
300 Washington Boulevard, Besson Hall, Fort Eustis, VA 23604, USA. http://www.transchool.lee.army.mil/museum/transportation%20museum/acv.htm

National Maritime Museum:
Romney Road, Greenwich, London SE10 9NF, UK. http://www.nmm.ac.uk

Deutsches Museum Munich:
Munchen, Germany

Hellenic Maritime Museum:
Akti Themistokleous, Greattida, 18537 Piraeus, Greece

Battleship G Averof Museum:
Piraeus Harbour (trireme anchorage)

There is a wealth of museums now worldwide that have preserved high-speed marine craft. A search on internet will give a long list for the potential visitor!

General Reference Materials

We present below some books and papers that the reader will find useful when researching HPMV whether for general information or going a little deeper. You will find additional references to theoretical papers and materials in our books on Air Cushion Craft [1-4] and WIG Craft [3-2].

Amphibious Operations in the 21st Century.
Electronic text available from US Marine Corps at: http://www.quantico.usmc.mil/MCBQ%20PAO%20Press%20Releases/090430%20CDI%20Docs/CDI_AmphibOps21stCent.pdf.

US Navy Amphibious Operations.
The main Navy site with access to LCAC data, up to date photos etc. from LCAC operations right around the world, current data on LCS: http://www.navy.mil/navydata/ships/amphibs/amphib.asp.

Written Course NAME 4177—The Practical Design of Advanced Marine Vehicles at:
http://www.mckesson.us/mckwiki/index.php?title=NAME_4177_-_The_Practical_Design_of_Advanced_Marine_Vehicles. This is the written version of the University of New Orleans course NAME 4177 at the School of Naval Architecture and Marine Engineering, College of Engineering. The course is a 13-week undergraduate elective course. The written material is very useful as a summary of HPMV design with focus on Catamarans, SWATH and SES.

How to Design a Boat.
By John Teale, 3rd Edition, 2003, ISBN 978-0-7136-7572-6, Adlard Coles, London W1D 3QY, UK. A practical designers approach to small craft design following traditional naval architects practice, including fast planing craft. Go to http://www.adlardcoles.com.

International Code of Safety for High Speed Craft (HSC Code).
International Maritime Organisation, IMO Publication 187E, first published 1994, with updates 2000, 2004, and 2008, ISBN 92-801-1326-7. Search on http://www.IMO.org, IMOdocs for updates and explanations, and also at http://www.KHUP.com with keywords HSC Code for extracts updates and referring papers.

Rules for the Classification of High Speed Light Craft and Naval Service Craft.
Det Norske Veritas, Veritasveien 1, 1322 Høvik, Norway, ref NV.1.85.3000, electronic version available for download at http://www.dnv.com current version January 2011, updates are issued regularly by DnV.

Rule and Regulations for the Classification of Special Service Craft, 2010.
Lloyds Register, London, Volumes 1–8, free pdf download at http://www.webstore.lr.org under Marine Downloads.

Guide for Building and Classing High Speed Craft, 2001.
American Bureau of Shipping, 16855 Northchase Drive, Houston, TX 77060, USA, downloads available at http://www.eagle.org including supplements and commentaries up to November 2010.

The Ship: An Illustrated History.
By Bjorn Landstrom, Doubleday & Co Inc, 1961, ASIN B0006AX9Z2.

British Motor Gun Boat 1939–45.
By Angus Konstam and Tony Bryan, Osprey Publishing Ltd (http://www.Ospreypublishing.com), 2010, ISBN 978 1 84908 0774/0781, 48 pages.

Assault Landing Craft: Design, Construction and Operations.
By Brian Lavery, Seaforth Publishing, 2009, ISBN 978-1848320505, 120 pages.

Dynamics and Hydrodynamics of Surface Effect Ships.
Kaplan P, Bentson J, Davis S, SNAME Annual Meeting Papers November 1981.

Operational Characteristics Comparison (ACV and SES).
Wilson FW, Viars PR, Paper AIAA-81-2064 presented at AIAA 6th Marine Systems Conference, September 1981.

Bell Halter Surface Effect Ship Development.
Chaplin JB, Paper AIAA-81-2072 presented at AIAA 6th Marine Systems Conference, September 1981.

Dynamics of SES Bow Seal Fingers.
Malakhoff A, Davis S, paper AIAA-81-2087 presented at AIAA 6th Marine Systems Conference, September 1981.

The Surface Effect Catamaran—A Sea Capable Small Ship.
Wilson FW, Viars PR, Paper AIAA-81-2076 presented at AIAA 6th Marine Systems Conference, September 1981.

Fast Passenger Ferries and their Future.
Wang J, McOwan S, Maritime Policy and Management, Volume 27, Issue 3, 2000, pages 231–252. DOI 10.1080/030888300411086.

The Quest for Speed at Sea.
Clark DJ, Ellesworth WM, Meyer JR, Technical Digest, Apr 2004, 25 pp, Naval Surface Warfare Center, Carderock Division, available for viewing at http://www.Foils.org under Hydrofoil technical references.

Wing in Ground Effect Vehicles.
Rozhdestvensky KV, Progress in Aerospace Sciences, Volume 42, Issue 3, May 2006, pages 211–283, published by Elsevier, available electronically via ScienceDirect. A survey of WIG vehicle projects, science and certification as it stood in 2006.

An Investigation into Wing-in-Ground Effect Airfoil Geometry.
Moore N, Wilson PA, Peters AJ, School of Engineering sciences, University of Southampton, UK, presents wind tunnel comparative tests of DHMTU and NACA series airfoils in 2002–2003.

Revisiting artificial air cavity concept for high-speed craft.
Gokcay S, Odabasi AY, Insel M, Ocean Engineering 31 (2004) pp 253–267, available at http://www.sciencedirect.com. A review and model testing for improved performance of air cavity craft performed by personnel from the Department of Naval Architecture of Istanbul University.

Application of Air Cavities on High Speed Ships in Russia.
Sverchkov AV. Krylov Shipbuilding Research Institute, St. Petersburg, Russia. International Conference on Ship Drag Reduction, SMOOTH-Ships, Istanbul, Turkey, 20–21 May 2010. This paper details the development of craft such as Linda, Serna, Saigak, and Merkury.

Chinese Language Sources

"High performance marine vessels"
Zheng M et al: National Defense Industry Press, Beijing, China, 2005 (in Chinese)

"The principles and design of high performance ships"
Zhao LE: National Defense Industry Press, Beijing, China, Jan 2009 (in Chinese)

"Hydrodynamic of high speed craft"
Dong ZS: Navy Engineering Academy of PLA, China, 1985, (in Chinese)

"High speed marine vessels"
Liang J et al

"Images for high speed marine vessels"
Lu PX et al

Index

MIX
Papier aus verantwortungsvollen Quellen
Paper from responsible sources
FSC® C105338

If you have any concerns about our products,
you can contact us on
ProductSafety@springernature.com

In case Publisher is established outside the EU,
the EU authorized representative is:
Springer Nature Customer Service Center GmbH
Europaplatz 3, 69115 Heidelberg, Germany

Printed by Libri Plureos GmbH
in Hamburg, Germany